AUTOMATION IS A MYTH

AUTOMATION

IS A MYTH

LUKE MUNN

STANFORD UNIVERSITY PRESS
Stanford, California

STANFORD UNIVERSITY PRESS
Stanford, California

©2022 by Luke Munn. All rights reserved.

No part of this book may be reproduced or transmitted in any form or by any means, electronic or mechanical, including photocopying and recording, or in any information storage or retrieval system without the prior written permission of Stanford University Press.

Printed in the United States of America on acid-free, archival-quality paper

Library of Congress Cataloging-in-Publication Data

Names: Munn, Luke, author.

Title: Automation is a myth / Luke Munn.

Description: Stanford, California : Stanford University Press, 2022. | Includes bibliographical references and index. |

Identifiers: LCCN 2021034838 (print) | LCCN 2021034839 (ebook) | ISBN 9781503631113 (cloth) | ISBN 9781503631427 (paperback) | ISBN 9781503631434 (ebook)

Subjects: LCSH: Labor supply—Effect of automation on. | Employees—Effect of technological innovations on. | Automation—Social aspects.

Classification: LCC HD6331 .M8865 2022 (print) | LCC HD6331 (ebook) | DDC 331.12/5—dc23

LC record available at https://lccn.loc.gov/2021034838

LC ebook record available at https://lccn.loc.gov/2021034839

Cover design: Kevin Barrett Kane

Text design: Kevin Barrett Kane

Typeset at Stanford University Press in Minion Pro

Contents

AUTOMATION IS A MYTH

INTRODUCTION

Automation Is a Myth

AT A LUXURY SKI RESORT in the heart of Switzerland, the snow has just started to fall, big flakes drift-ing down through the thin mountain air. An elite group of global leaders—presidents, heads of state, policymakers, CEOs, busi-ness magnates, media personalities, celebrities—had descended on the alpine village over the past few days. Now they were filing into the conference center, quickly snapping up the last empty seats in the packed auditorium. After a brief introduction, the house lights dimmed and a bank of massive screens behind the stage lit up. As a figure walked onto the stage, the title of his talk, "How to Survive the 21st Century," flickered onto the screens. The "automation revolution," he proclaimed, "will be a cascade of ever bigger disruptions. Old jobs will disappear, new jobs will emerge—but then the new jobs will rapidly change and vanish." A Windows-style error message now appeared on-screen with a bright yellow warning sign: "Technology might disrupt society and the meaning of life itself," it cautioned, offering Ignore or Reset buttons. With a group of global elites hanging on every word, the speaker ramped up his rhetoric and rammed home one of his takeaway points. "Whereas in the past, humans had to struggle against exploitation," he warned, "in the 21st century, the really big struggle will be against irrelevance."[1]

Race Against the Machine. A World Without Work. Rise of the Robots. The Future Is Faster Than You Think. The Inevitable. From industry forums to technology journalism and the front pages of the popular press, automation rhetoric abounds, and for good reason. This is a tale of drama, where technologies suddenly sweep in and society undergoes massive upheaval. This is a tale of crisis, where work as our long-standing means of identity and economic support is now placed in jeopardy. And this is a tale that is universal, obliterating geographical and cultural distinctions to insist that automation will affect everyone, everywhere. "It doesn't matter what you do for work," asserted one pundit; in a matter of years, your job will not be the same.[2] "Automation is the terrifying force no one is willing to name," argued another, "the race between automation and human work is won by automation."[3] These stories simultaneously hold out the promise of economic prosperity while warning about a set of potentially disastrous effects. They form a powerful narrative driven by hype and fear.

Automation is a myth, a long-running fable about the future of work that needs to be reconsidered. Whether embraced as dream or cautioned as nightmare, automation is ultimately a fiction, a fantasy. "Myth" does not imply that automated technologies do not exist or that there have not been technically driven transformations in the nature of work over the past century. But these transformations have been piecemeal rather than total. They have taken place differently within different cultures and locations. And they have impacted particular races and genders rather than a generic humanity. In speaking of "myth," then, I'm taking aim at the universal understanding of automation constructed by population automation discourse, a fable that abounds in press headlines and popular best sellers. Indeed, this version of automation must be a fable because it rests on a set of triple fictions: the myth of autonomy, the myth of automation everywhere, and the myth of automating everyone. These claims are disconnected from the sociomaterial conditions of the real world, divorced from the lived experiences of specific bodies in specific spaces. As a result, this framing of automation is a hollow concept, resting on nothing. Untethered from the hard ground of reality, it lacks any capacity to reveal the power relations, technical operations, and logistical chains that contour contemporary labor. "Automation" signifies so much but explains so little.

Why does this myth matter? In grabbing headlines and framing the debate, this myth is not just misguided but dangerous. The automation myth offers a cohesive narrative, a "unified vision" for a set of conditions that are actually fragmented and uneven, and this myth exerts persuasive force, working "to explain, to reconcile, to guide action or to legitimate."[4] In claiming to be apolitical, automation camouflages its ethics. This fable about the future of work often props up existing hegemonies, reinforcing relationships of domination when it comes to technology, labor, race, and gender. In claiming to be global, automation overlooks its context. Its abstracted, 10,000-foot view obscures the small but significant transformations taking place in that factory in that city—material shifts that reconfigure labor conditions for particular people in particular places. And in claiming to be inevitable, automation conceals its interests. Its vision of a ubiquitous, impersonal force overlooks how change really occurs—through identifiable actors, with certain motivations, making concrete decisions.

In accepting the framing of automation, we not only miss the key questions but the chance to intervene within these conditions and shape these outcomes. Indeed, perhaps the most damaging aspect of the automation myth is its fatalism. Much of what will happen is "inevitable," we are told, driven by "technological trends that are already in motion."[5] Here, history becomes a tale of technical improvement with power, speed, standardization, uniformity, and control as the given variables for optimization. Nothing can alter these axioms or stop this forward progress. In the myth of the machine, the machine is "absolutely irresistible."[6] The future of work is a future on rails. To reject the myth of automation, then, is to reject this fated future and insist on alternatives. Other variations and inflections—more communal, more equal, more sustainable—are possible.

All of these blind spots prevent us from properly perceiving new entanglements of machinic and human labor. For instance, I could tell a very different story about automation. "I'm a shared-bike redistributor," states the man, a heavy-set youth with a goatee dressed in an old Guess T-shirt. Puffing on a cigarette, he sizes up a jumbled mess of bikes heaped on the street, then starts to pull them apart. "I work through the night, putting these shared Meituan bicycles back in places where they'll be needed at rush hour." Grunting, he grabs the frame of a bicycle, hauling it onto a trailer attached to a three-wheeled cargo motorcycle. One by one, he finds and loads bikes, working up a sweat despite the cool night air. "People attribute my work to the magic of an app," he notes, "but these bikes don't teleport themselves." He scans the street for a final bike to complete the load—then groans when he spots one dangling halfway up a tree, wedged between two branches. By standing on a concrete planter, he can just reach the front wheel, gripping the tire and levering it out of the crook in the tree. Slowly lowering it to the ground, he feels a twinge in his back but presses on, completing the load and firing up his cargo bike. "I guess I'm a fauxtamaton," he admits, "invisible labor that keeps up the facade of shiny technological efficiency." Reaching the inner city, he cuts the engine and starts to unload each bike, setting them up in a neat row ready for early-morning commuters. With a grunt, he grabs the final bike with one hand and swings it over his shoulder, grinning briefly: "I'd like to see an algorithm try this."[7]

This is just one story, about one person, in one place. Yet its specificity stands in sharp contrast to the grand rhetoric of "humanity" and "technology" in the opening scene. It points to a set of particular pressures placed on the body by technically mediated labor. It highlights some of the potentials and punishments enabled by digital affordances. And it gestures to a more nuanced set of social and political issues. It suggests that working through this myth would be productive, that questioning its universal categories and critiquing its seemingly inevitable trajectory could be beneficial, providing us with more articulated insights about automated technologies and the future(s) of work.

Materials and Lenses

To clear away this myth and replace it with a sharper portrait, this book draws together a diverse variety of materials, an eclectic mix that aims to offer a cross-section of different labor forms. Warehouse work is just one of these forms, but it offers a prime example of what material was chosen, how it was approached, and what this engagement aimed to accomplish. With the rise and rise of e-commerce, these sites have become key distribution centers, coordinating a massive performance that delivers millions of goods every day. Speed and efficiency are the prime directives here, and so the warehouse has long been a field site for the future of work, a test bed for new technologies seeking to rationalize and routinize labor. The dream is to perform "like the motherboard of a computer, combining and redistributing goods as bytes and containers like software containers."[8] Automation aims to smooth out the lumpy tasks of packing and stacking into a seamless logistical flow. Yet this choreography remains frustratingly complex and human bodies frustratingly difficult to erase. This makes the warehouse a key site of struggle, a space that crystallizes the tensions and limitations of automation.

To develop a portrait of these conditions, I synthesize a wide array of "gray" literature, from journalistic investigations to computer science papers, patents, business reports, and worker testimony. I want to zoom in on the point where the technical butts up against the social and material, examining the interplay among automated systems, automated spaces, and automated subjects. I aim to freeze-frame those moments when the world bites back in the form of error, injury, and contingency. Of course, the warehouse is just one workplace of many. To augment this portrait, the book ventures into other sites, from the factory to the field and from the supermarket to the smart home. Together these aim to touch on the broad spectrum of practices and conditions that fall under that enormous umbrella we call work.

How do we interpret this material and make sense of these conditions? Here I turn to media, race, gender, and cultural studies and the deep insights they offer. These scholars throw a spanner in the machinery of the automation myth, bringing its grand claims to a temporary halt. They urge us to pause, to carefully inspect relations and conditions, and to pay attention to adverse impacts, pointing out that this is "not a mere technical emergence but the practical result of an ongoing and bloody struggle between the would-have-it-alls and the to-be-dispossessed."[9] Automation privileges some and punishes others. Such scholarly work grounds automation in its colonial history, highlights its racial and gender inequalities, and stresses its political agency. By reading this material through a critical lens, this book aims to offer a more nuanced, localized, and racialized understanding of automated technologies and their intersections with labor regimes. Specific lives and the spaces they inhabit provide an anchor, bringing automation back down to earth.

Fiction by Fiction

This book steps through each of automation's fictions in turn. Part 1 examines the myth of *automated autonomy*. For decades automation rhetoric has assured us that total automation is imminent, that machines will soon operate completely independently, taking over production and rendering the human obsolete. But work has proven to consist of a range of non-trivial problems, full of inconsistencies and edge cases. Increasingly it seems that the horizon of full automation will never be attained. Instead we see a revision of this dream—a more modest set of technical interventions that actively acknowledge humans and our rich capabilities. Far from being self-acting, these technical systems require heavy amounts of development and maintenance. Alongside this highly paid work is precarious and exploitative piecework carried out by a digital underclass. These insights highlight the immense amount of human labor behind "autonomous" processes. Automation is incomplete, and Chapter 2 moves from checkout operators to machine minders and content moderators to explore what this partially automated labor looks and feels like.

Part 2 investigates the myth of *automation everywhere*. Automated technologies are frequently framed as a wave or an age, a de-situated force that will sweep across society or ripple across the globe. But this fiction ignores the social, cultural, and geographical forces that shape technologies at a local level. Automation is both technical and geopolitical, and any discussion must situate the impacts of these technologies within a specific context. To highlight this point, these chapters move through examples of automation in China, jumping from *shanzhai* practices in Shenzhen to logistics in Hangzhou. Each demonstrates how technologies emerge from domestic ecosystems, reflecting the distinct values and visions of the cultural landscape that surrounds them. The second chapter in this part dives deep into two instances of automation in Xinjiang to demonstrate how a cultural and contextual understanding is key. To grasp what automation is doing in these cases, we need to go beyond the technical and draw on the historical, social, and racial dynamics at work in a particular place.

Part 3 explores the myth of *automating everyone*, the generic figure of "the human" at the heart of automation claims. Throughout the decades, automation discourse has been dominated by terms like "humanity" and "mankind." Automation would affect us all equally. But this framing obscures the fact that labor is socially stratified along racial and gendered lines. Automation emerges from a colonial history that valued some humans more highly than others, and automated technologies are deployed in industries that are already structurally unequal. This means that automation's fallout will also be uneven, falling more heavily on some than others. By zooming into a single warehouse, we can see how these transformations particularly attack Black lives, producing injuries and terminations but also fostering forms of activism. Chapter 6 turns to gender, a shift that highlights how automation's definition of work—waged work in the workplace—is narrow and patriarchal, dismissing an entire realm of labor and the contributions of those who undertake it. The chapter pivots from industry to domesticity to challenge this blinkered concept, investigating the non-automation of "non-work," the masculine vision of home automation, and the feminization of digital assistants.

Together, these stories push back against the myth of automation, using it as a kind of springboard toward a sharper articulation. By examining conditions on the ground, we can develop a more fine grained portrait of the technical reshaping of work. The aim here is not to add more detail for the sake of it, but to offer a conceptualization able to account for particular forces exerted over particular subjects within particular spaces. The interplay that emerges is messier but more interesting. Here we find control but also contingency, we see cultural logics that dictate technical logics, and we witness racial "histories" that persist in the present. These forces deeply shape the power relations surrounding labor: the way that work is carried out, the agency that workers have, their ability to sustain their livelihoods, and the forms of community that can be fostered. In other words, automation is not just technical but political. By sharpening our understanding of automation, we sharpen our understanding of its stakes and table a sharper set of claims. To insist that certain groups are accounted for, that certain values are upheld, that certain configurations of technology and labor must be entirely redesigned—these are specific demands that require reimagining form and function in concrete ways. "Automation"—that myth claiming to narrate the future of work for all people in all places—needs to be rewritten.

PART 1

The Myth of Automated Autonomy

1

The Fantasy of Full Automation

AUTOMATION IS A MYTH because its central conceit of humanless work is unattainable. At the root of the word *automation* is a claim to become self-acting, to move and act on its own,[1] and automation discourse continues to insist that this level of independence is right around the corner. In this alarming fable, machines—from robotics to software systems and algorithmic processes—are rising up. Better, faster, and more productive, they will soon perform all aspects of labor. "As they slowly, but relentlessly, take on more and more tasks," warns one economist, in a statement typical of the automation genre, "human beings will be forced to retreat to an ever-shrinking set of activities."[2] The human, we are assured, is on the way out.

The pursuit of full automation is a long-standing dream. One of the first patents filed with the newly established US Patent Office in 1785 was for an automated flour mill invented by Oliver Evans. By employing a complex series of levers and devices, Evans was able to reduce the number of laborers from four to just one. These machines, he explained, "perform every necessary movement of the grain and meal, from one part of the mill to another, and from one machine to another, through all the various operations . . . without the aid of manual labor, excepting to set the different machines in motion."[3] Already at this early stage, we can see technology chipping away at the dependence on "manual" labor and that most ancient of devices: the human body.

From here a genealogy of automation could take many forms, but we might focus on the factory and warehouse, picking out just a few moments. By 1920, the A. O. Smith Corporation had constructed an automated factory that performed 500 operations in order to produce over 10,000 car frames a day; the facility was considered a "wonder of its time and a prototype for the factory of the future."[4] In the mid-1950s, the first automated guided vehicle was introduced into warehouses, a driverless tractor train called the Guide-O-Matic that followed a wire embedded in the floor.[5] A decade later, automated storage and retrieval systems (AS/RS) debuted in the warehouse. In essence, these were computer-controlled stacker cranes that could shuttle up and down to pick up or deposit an item. Automatic retrieval allowed shelving to stretch to the ceiling and rows to become narrower, increasing the density of storage.[6]

This flurry of innovations contributed to a sense of inertia, fueling the belief that technology would soon unburden society of its toilsome labor. Indeed, this period, which loosely coincides with the Space Age, is marked above all by a profound awe of technology, a reverence for the immense progress that science seemed to be making in every sphere of life, including work. Given this advancement, it would be no time at all before automated systems would supersede humans altogether. "Until the middle of the twentieth century, a non-aggression pact stabilized a more or less peaceful coexistence between humans and machines," warned Georges Elgozy in 1968. "Infinitely more powerful and autonomous, the automations of the second half of the twentieth century will quickly break this equilibrium. Item by item, they will replace their less-evolved ancestors."[7] The interventions of the past were trivial, leaving existing work relations largely intact. But the inventions of the near future would soon result in total autonomy, with machines achieving full spectrum dominance.

In the 1980s we can see one pinnacle of this dream in the form of the "lights-out" factory. General Motors had once been a great carmaker, a titan of the Detroit automotive industry. But over the years, its market share had been rapidly slipping to Japanese competitors, who were able to produce vehicles faster and cheaper than their American counterparts.[8] For management, a fully automated facility was the clear solution, a way to claw back control and reassert their dominance. Every aspect of production would be thoroughly mechanized, with vehicles assembled from start to finish by machines. And every single slow, error-prone, and sometimes antagonistic human worker could be jettisoned, saving millions of dollars in labor costs per year. Once the factory had attained this lofty peak of efficiency, the last symbolic concession to the human—light to work by—could finally be removed. In this dark and cavernous space, machines would work silently and incessantly to churn out perfect products in record time. The company would be saved; shares would skyrocket. In fact, GM's dream would ultimately fail. But if that incarnation faltered, the lights-out factory remains a compelling dream for many in the industry, maintained in large part by the promises of always-improving, always-accelerating technology.

Always Improving, Always Accelerating

Alongside this history of technical innovation is a history of computational acceleration that automation discourse loves to highlight. Martin Ford devotes several pages in his best-selling *Rise of the Robots* to Moore's Law, Gordon Moore's 1965 prediction that the number of transistors on microchips would double every year. Erik Brynjolfsson and Andrew McAfee also draw on Moore's Law for a section of their popular book, *Race against the Machine*. They compare the law to the foundation myth of chess, where its inventor asks the king for a modest reward of rice for every square on the chessboard: one grain on the first square, two grains on the second square, four grains on the third square, and so on.[9] The authors of both books exert great effort to communicate the power of the doubling that results, drawing on analogies and illustrations. The exponential advancement of processing power boggles the mind, but readers must be made to understand.

In fact, already by 1975, Moore had to revise his prediction, adjusting it to be a doubling every eighteen to twenty-four months. And the "law" stopped holding altogether around 2010, as controlling the flow of current became increasingly difficult with the shrinking size of chips. But these are minor quibbles that ignore the crucial work that Moore's Law is doing in this discourse. In both books, it's apparent that something more is at stake than a mere doubling of transistors on a chip. The acceleration of microchips is linked to the overall advancement of technology, the ability to use this technology to carry out labor, and the result of a steady improvement in society. For Martin Ford, it's obvious that machines are moving from being tools to become workers in their own right and that "all this progress is, of course, being driven by the relentless acceleration in computer technology."[10] Here Gordon's modest technical trend begins to stand in for something much greater. Claims around processor power balloon into far grander claims about technical power and humanity's ability to control machines to carry out work. Moore's Law = progress. Every year, technology is not just getting faster, but also better and more sophisticated. Every year, our previous expectations are shattered, and the new normal gets revised. Given this exponential improvement, it is only a matter of time before we will witness "computers that can accomplish previously impossible tasks."[11]

Moore's Law and broader technology development help prop up the checkered history of automation. The compelling dream of full automation gets reenergized based on breathless claims about the advancements made by newer technologies.[12] It may not have worked then, the argument goes, but this time will be different. In our current incarnation, hopes and dreams are pinned on nebulous AI. Artificial intelligence, we are told, will change everything. Listen to highly influential tech leaders like Elon Musk, and you'll hear about the "alien dreadnought" factory, a facility asserting that the future of work will be "completely inhuman" by design.[13] Watch the promotional videos of technology titans like Alibaba and Amazon, and you'll be presented with a vision of smooth logistical flow devoid of human touch. In these visions, vast warehouses are populated by swarms of machines that ceaselessly create products and ship packages in a pixel-perfect choreography. Work is performed. No humans required.

If we believe this tale, the fate of the human becomes all too clear. Technological developments are typically rationalized in the blandly technocratic terms of efficiency, cost savings, and accuracy, slotting neatly into broader industry rhetoric of innovation and modernization. And yet as scholars like David Noble and Harry Braverman have thoroughly demonstrated, the history of automation is not just technical but social in reconfiguring the role of the human worker.[14] Implicit in this history is a drive to reduce human labor: fewer employees, less time needed, fewer skills required. The end game of this dream is uncompromising: the production and distribution of goods will increasingly take place through mechanical or technical means, while the human laborer is first sidelined and then dispensed with altogether, banished from the world of work. Efficiency and eviction go hand in hand.

Unable to match the relentless pace and exacting precision of the machine, humans will become superfluous, a vestige of a former age. For some commentators, this freedom from labor would be welcome, ushering in a flourishing utopia,[15] a postcapitalist "world without work,"[16] or a form of "fully automated luxury communism."[17] These visions often draw on a strain of leftist and Marxist thought that has long considered machinic production to be a potential form of liberation, freeing workers up to enjoy leisure or more principled pursuits. For others, the fully automated future is a much bleaker prospect, validating a long line of warnings over the decades concerning "technological unemployment,"[18] the "displacement of men,"[19] the dystopia of mechanization,[20] the "automation jobless,"[21] and the automation crisis.[22] But while the automation optimists and the technopessimists may differ on the implications of this shift, they share the same underlying assumption of technical takeover. Whether embraced as a bright future or rejected as a dark dystopia, the prognosis is remarkably consistent: full automation will fully replace the human.

Non-Trivial Problems

On the face of it, this historical progression seems to imply a clear arc, a vector moving smoothly and inexorably toward a fully automated future where the human laborer is no longer needed. But accepting this narrative at face value would be a mistake. Here it pays to attend to a specific context such as the warehouse. For one, there are technical constraints. The identification, packaging, and redistribution of goods is a non-trivial problem. Certainly advancements in fields like machine learning have enabled new headway into this domain, with an array of research chipping away at technical challenges. And yet there are still tasks that remain almost effortless for humans while being highly difficult for machines.

As early as the '70s, one logistics expert for Kodak was criticizing the "overautomated warehouse."[23] A new automated crane system had been installed and needed to calculate the shortest routes for picking up items around the warehouse. Yet this computation was so complicated that it produced long delays, barely offering an advantage over a human crane operator in terms of speed. The warehouse, in fact, had hit up against the infamous "traveling salesman" problem, a so-called NP-hard problem that is computationally difficult. Far from disappearing, this problem has only been exacerbated in the modern warehouse, with its millions of products and thousands of orders waiting to be fulfilled. In a 2020 paper by the Chinese logistics giant Cainiao, the engineers lament the "huge solution space" presented by this problem and devise complicated frameworks in an attempt to "reduce the computational complexity and meet the requirements of time efficiency."[24] These are hard problems with no simple solutions.

In the previous section, we saw how automation pundits use Moore's Law as a kind of proof of steady advancement. A trend in transistor density becomes a far grander claim of technical mastery over time. Every day, in every way, technology is getting better. But computing power is not everything. For proof, we only need to turn to the modern field of machine learning, where data scientists aim to generate analytical models automatically from data sets. In some ways, machine learning is the poster child for processor acceleration: theorized seventy years ago, it remained relatively abstract and underdeveloped until chip improvements in the late '90s started to render it practical as a technique. In the past decade, then, there is no doubt that machine learning has flourished, leading to new applications. Yet despite these gains, there are signs that certain abilities will not and cannot be achieved by merely ramping up the size of models. In the end, the "bigger is better" mantra of today's billion-node neural networks may yield only incremental improvements rather than the kind of novel learning that is actually desired.[25] Indeed, some critics have suggested machine learning is reaching its limits, that its capabilities have begun to plateau and that another "AI Winter" is approaching.[26]

Alongside technical constraints, there are financial, material, and organizational constraints. The lights-out vision championed by General Motors, as alluded to, was generally considered to be a failure. Rather than the smoothly routinized process that was envisioned, automation caused chaos in the factory. In their book on the American auto industry, journalists Paul Ingrassia and Paul White lay out the litany of problems that resulted.[27] The computer-controlled dolly wandered off course. Spray-painting machines painted each other instead of the cars, forcing GM to send cars across town to an old Cadillac plant for painting. Robots were unable to tell cars apart, mistakenly installing Buick Riviera bumpers on Cadillac Sevilles. And when a robotic welding machine crashed into an auto body, the whole assembly line ground to a halt. Workers would have to stand around while they waited for a technician to arrive. Even after investing billions in the scheme, productivity at some plants actually went down, and the Japanese cost advantage remained unchanged. The "great automation solution" had fizzled and flopped.[28]

The same disappointments have more recently plagued the Tesla factory. "Excessive automation at Tesla was a mistake," admitted Musk in a much publicized tweet; "humans are underrated."[29] A year earlier, the CEO had announced his ambitious plans to produce a fully automated factory. And yet implementing this vision had resulted in serious problems. Robotic systems had been unable to adapt to the inconsistencies in assembly tasks, putting the company far behind its production targets. After coming to terms with this failure, the strategy was rethought and the facility reconfigured to integrate human labor in a more traditional way. The integration of robotics into a production or warehousing environment, as even optimists like Brynjolfsson and Beane admit, has proven to be a complex and highly expensive undertaking, with many companies instead opting for much more modest plug-and-play automation they can integrate into existing systems and processes.[30]

Routinizing work is not easy. Machines generally require regularity and uniformity. But labor, it turns out, is highly complex and highly differentiated. Tasks might require a huge variety of hand gestures, or a deep knowledge of the domain to make decisions, or the ability to improvise and deal with crises—all areas where machines struggle and humans excel. Transforming these diverse tasks into a process that is repeatable, predictable, and therefore machine operable is easier said than done. When we think of the sheer number of industries spread across countless countries, each with its own complexities, its own contexts, and its own ways of operating, we can begin to get a sense of the sheer number of site-specific applications needed to "solve" the problem of work. Indeed it becomes clear that the simple word "work" disguises a sprawling terrain of wildly divergent environments, processes, and people.

To see these limits in action, one only has to think about less standardized forms of work. Personal services, for example, is one of the fastest-growing job markets today. From animal carers to child care workers, fitness instructors, and home health aides who assist the elderly or chronically ill, these jobs are deeply situated. Workers generally develop a relationship over time with a person or animal and draw on this deep knowledge to attend to their specific needs. To labor in this context might mean fetching groceries and filling medical prescriptions. It could entail giving a bath while avoiding scrubbing bruises. Or it might require discussing memories from the war, developing a child's motor skills, or simply holding a hand. As Jason Smith notes, "the types of labor processes many automation theorists suggest are vulnerable to replacement by smart machines in fact require an intuitive, embodied, and socially mediated form of knowledge or skill that even the most advanced machine-learning programs cannot master."[31] Automating one-on-one care is far from trivial.

From a technical standpoint, then, the vast landscape of work is dotted with land mines: gotchas and gray areas, niche applications and edge-cases. The mantra that new technology will resolve these issues has been repeated until it is no longer believed. Trust has been eroded by too many empty promises. "Full automation is not looking promising at all," stated one engineer; "the idea of fully eliminating workers hasn't been practical in the past 70 years, and there have been hundreds of prototypes of full automation that have never been commercialized."[32] Such confessions are all the more powerful in that they come from technologists rather than sociologists, from industry insiders rather than critical outsiders. "We have 10% left to go, but it now looks at least possible that the last 10% is 90% of the effort, and that we might need something different," observed one influential technology analyst; the industry is gradually realizing that the road map "runs out short of the destination."[33] Technical advancement, once seen as inevitable, has become stalled at key junctures.

The Fantasy of Full Automation

What can we learn from this lineage of disappointments? Aaron Benanav once stated that the specter of "full automation can appear as both a dream and a nightmare."[34] But this history suggests that full automation itself was always a fantasy, a vision decoupled from the complexities and contingencies of reality. Across the decades, automation reports have been dominated by a kind of armchair economics. In these sweeping overviews, trends were often extrapolated based on high-level national or international labor statistics over the years. Many of these predictions were accepted as gospel. But these analysts never put on a hard hat and ventured onto the worksites they were generalizing about. There were too many assumptions divorced from conditions on the ground, too much hand waving and hoping based on technical improvements.

This means that when we search for the locus of automation, for that imagined point where autonomous machines extinguish the human, we come up empty-handed. The fabled site does not exist. Instead we encounter, again and again, the fleshiness of human labor. On a warehouse floor in Hangzhou, a fleet of automated guided vehicles dart back and forth, each shuttling a package to a chute and dropping it in before returning to grab another. The vast facility, run by the Alibaba logistics arm Cainiao, seems to be empty save for these robots. "Packages are moving around freely in the warehouse with no people present," states the video's description.[35] Clearly an example of automation, the promo proclaims.

And yet once we start unpacking this scenario, the human reemerges and grand claims of autonomy begin to crumble. Follow the packages down the chute, and you'll find laborers collecting them on the floor below. These are bundled into paper bags and hauled into the back of a truck for shipping. Last year, just before the biggest shopping day of the year, logistics companies in China faced a formidable challenge: its couriers were not showing up to work. In a country where strikes are officially illegal under the law, companies like YTO Express and Yunda Express began reporting "abnormal operations."[36] The mostly young male drivers were walking off the job, protesting unfair wages and harsh work conditions. Express deliveries in Hunan, Jiangsu, and

Shanghai were affected and packages were delayed. This kind of improvised "non-strike" would hardly have been effective if human labor was as superfluous as automation discourse makes it out to be.

Revising the Dream

In recognition of these limitations, the vision of total automation is beginning to be revised. Companies have not only scaled back investments in automation, but more fundamentally rethought how this transformation should take place. "The fully automated or highly automated fulfillment center isn't a North Star we're trying to hit," declared one Amazon spokesperson.[37] There is a more open recognition of the limitations of automated systems. If the earlier vision of full automation presumed that machines could carry out work faster, better, and more efficiently, this amended imaginary seems more willing to acknowledge the constraints of computational awareness, autonomy software, and cyber-physical work systems.[38] The literature is marked by less hubris and more humility—although these terms would never appear in engineering articles or industry media outlets.

One result of this revision is a new turn to the human. Scan academic and industry papers, and you'll see terms like "human-centric" spike over the past few years. Instead of assuming a replacement of human labor, these papers speak of "symbiosis," discuss the "trade-offs between manual and automated processes," and probe what it might mean to have "shared responsibility" for certain tasks. This form of automation is premised more explicitly on collaboration between human and machine. Of course, the clear-cut boundaries between the anthropological and the technological presumed by that statement can certainly be contested. But this turn proceeds outside such theoretical debates, remaining uninterested in cyborgian figures[39] or the silicon-flesh hybrids fantasized about in transhumanism.[40]

At the same time, this industry-driven vision *is* highly interested in how bodies and systems might practically intersect. On an immediate level, there is a drive to bring bodies and machines into closer proximity, to remove the barriers that separate them. Brutally fast yet potentially dangerous industrial robots have been replaced in some settings by so-called collaborative robots. These "cobots" often feature in-built safety mechanisms that sense humans nearby and soft body parts that won't injure workers. In the same way, high-speed vehicles in warehouses—think here of Amazon's well-known floor bots that ferry goods—have been rethought or at least supplemented. One start-up came out of stealth mode several years ago to announce its "autonomous cart."[41] Drawing on the same kinds of sensors as self-driving cars, the cart senses humans and weaves around them. This is a slower but more human-friendly device, designed to slot seamlessly into the existing labor practices of warehouses and factories. The start-up has already been snapped up by Amazon, and so we can anticipate these kinds of devices rolling out into real-world spaces at scale.[42]

Alongside these specific examples, there is the more fundamental question of human-machine integration. Looking across industry portals and engineering articles, one can witness a push to more fully understand the touchpoints at play in these interactions. Such trade-offs are often not immediately apparent. Offloading physically demanding work to a cobot, for instance, might appear to be an obvious benefit. Yet this might mean that the natural rhythm of labor a worker once enjoyed—periods of physical exertion interspersed with brief moments of decision making—becomes lopsided. The worker's role now becomes purely cognitive—critical choices made at pace to match the machine—overloading their abilities or ultimately alienating them from the outcome. These are complex negotiations, then, that technologists and industries are still working through. The aim is to arrive at an optimal interplay so that automated labor becomes harmonious.

And yet there is much ground to make up. Automation's long obsession with efficiency has meant that it is somewhat unequipped to deal with the new shift toward meaningful human integration. Even those from disciplines like logistics and business management—well removed from the "softer" humanities and social sciences—have observed that the historical sidelining of the human has created a gap in knowledge. One supply chain expert admitted that the "literature is largely centered on design and technical factors related to performance, at the expense of human factors" and that even "the scant studies addressing human factors have mainly been from an ergonomics and safety point of view, neglecting socio-technical aspects."[43] This literature demonstrates how thin the understanding of the human within automation has been—a vestigial figure who only needs to be kept safe while the "real work" of technical advancement forges ahead. Who is the human in automation? What do workers need, what do they want, and what can they offer? Strangely these questions have only begun to be asked recently, allowing the "human-centric warehouse" to be presented as a "new paradigm."[44]

The new focus on humanly integrated automation, then, is a kind of doubling back. Earlier efforts at automation, moving from early industrialism to more advanced Taylorism, had a very diminished view of the human.[45] For some tasks, Taylor openly stated his requirements: a worker so "stupid" that "he more nearly resembles in his mental make-up the ox."[46] For other systems designers, the disadvantages of "human material" were numerous: they were "subject to fatigue, obsolescence, disease and even death"; they were "frequently stupid, unreliable and limited in memory capacity."[47] The ideal human was someone who carried out the given instructions to the letter, hour after hour, at pace. In these frameworks, the worker was barely more than a mechanism that lifted or pushed or swiveled—man was a motor function. If a mechanical system could consistently and accurately replicate these basic physical motions, it could effectively replace the human.

Yet if these interventions were certainly productive up to a point, they failed to recognize (and instrumentalize) the full extent of cognitive, psychological, and social capacities possessed by the human. Things were missed in the single-minded pursuit of totalized automation. In the warehouse, for instance, these decidedly human skills might include the ability to identify an item buried underneath other products or to improvise on the fly when a product cannot be found in the correct location. In content moderation, to take another example, the human ability to understand the subtle social and cultural context of messages is highly valuable. These skills are deceptively simple for humans but have proven frustratingly difficult to code and computerize. In revising the dream of full automation, there is a renewed recognition that the human was a far richer subject than was ever supposed.

Entangled not Erased

In *Our Robots, Ourselves: Robotics and the Myths of Autonomy*, technologist David Mindell underscores many of these same points, pushing back against the inevitability of full automation. The concept of linear progress assumed that technology would constantly improve and that human presence in the labor sphere was fated to fade away, shifting from direct involvement to a remote or mediated intervention before autonomous systems supplanted them altogether.[48] According to this fable, humans would soon be "out of the loop" altogether. But there is no evidence to suggest that this is the case. In fact, counters Mindell, the opposite is true: the future suggests a closer coupling between technical systems and human workers, a deeper intimacy between people and their machines. "For any apparently autonomous system," notes Mindell, "we can always find the wrapper of human control that makes it useful and returns meaningful data."[49]

We see instead a renewed push to actively integrate human labor into automated processes. This is not simply a matter of using human labor for small tasks unable to be automated away, but instead a deeper enmeshment of machinic and anthropocentric work. Turning once more to the warehouse sector provides a glimpse of what these developments look like in practice. One manager of an Amazon fulfillment center said they are striving to make it "feel seamless between what the robot is doing and what the humans are doing."[50] Another stated they're able to "leverage the things that people are good for and leverage the things that systems are good for. We see that pattern playing out in a lot of different applications."[51] What role should each play in this new labor regime? How might the interplay of decisions and gestures best function? And what adaptations in automated systems are necessary to bring this synergy about? These are live questions that are being worked out on the floors of worksites around the world.

Technical systems and human labor are deeply entangled, a relation that Louise Amoore highlights in *Cloud Ethics*. While Amoore's case study of surgical robots is certainly distinct from warehouses, she stresses that here too, rather than the "evacuation" of the human that was forewarned, the surgeons describe an "extension" of their self and their capabilities as machinic vision and devices open up new possibilities for seeing and sensing. The interaction between these machines and their users is collaborative, a deeply "reciprocal and iterative relationship" where users adapt to technical constraints and possibilities, and the machine records and learns from the user.[52] The movements of surgeons become training data for these systems, allowing the machine to "learn" to stitch based on thousands of collected examples. In this sense, the optimal machinic cut offered by the robot congeals the labor of thousands of surgical teams over several years. "In every singular action of an apparently autonomous system," Amoore stresses, "resides a multiplicity of human and algorithmic judgments, assumptions, thresholds, and probabilities."[53] These insights remind us that even when machines appear to be "self-acting," they are deeply infused with human intelligence and expertise. Data sets and programming are dense packages of human labor, collected over years and crystallized into digital forms.

This entanglement draws our attention to new forms of subjectivity that automated technologies introduce. The adoption of new technology in the workplace is not simply a means of doing the same thing faster or better, but fundamentally shifts a worker's agency, experiences, and expectations. Some might see these shifts as abhorrent; others might consider them exciting or simply necessary. I'm less interested here in moral condemnation and value judgments and more interested in considering this technical condition as an imperative that confronts the laborer. For better or worse, this is the new normal that must be accommodated and negotiated. Indeed, technically mediated labor fundamentally reworks such values, reconfiguring the very notion of what it means to be a "good worker," a "team player" who performs effectively and identifies positively with her labor. Immersed in these conditions, how does the individual "invent a self-understanding that optimizes or facilitates their participation in digital milieus and speeds"?[54]

In this context, the Amazon warehouse worker—or "associate" in the company's parlance—is the canary in the digital coal mine. Linked to a wearable device and a range of sensors, her steps, item scan count, and even bathroom breaks are tracked and used to generate key performance indicators. This is a work performance that is exhaustively digitized, and so it is not enough to do the work; her work must be made machine-legible. How is work categorized within this software, and what tasks are prioritized? Which variables are indexed, which ones matter, and which ones are shared with management? Gradually the worker gets a feel for this technically mediated work, a tacit grasp for how certain activities will be interpreted. Over time, workers internalize the logic of the system and perform their activities in a way that is algorithmically recognized. To see this in crass terms as gaming the system is to miss the more subtle subjectivation taking place here. This is a mutual adaptation, as the worker accommodates herself to technical systems and these systems "learn" from the worker. Of course, these systems do not run themselves or always run flawlessly, and the next section investigates the human labor that props them up.

Maintenance, Not Magic

Watch the promotional videos produced by automation companies, and you'll be presented with a flawless universe. In the pixels of this CGI world, systems function perfectly and hardware remains forever glossy. Here, moth and rust do not destroy. Yet in the real world, things break. Plastic cracks, metal warps, chips overheat. Technology needs maintenance, not magic. To labor in this space might mean monitoring systems, diagnosing problems, swapping out parts or code, installing new components and adapting them to a specific context or application. This is not sexy work. For server techs, for instance, a "truck roll" might mean driving to a data center in an industrial park at midnight, parking, dragging a keyboard and mouse over to a server, identifying the problem, replacing a dead hard drive, and restarting, before heading back home. These are the unglamorous jobs, the roles never featured in starry-eyed tech reportage focused on founders and innovators.

In automation discourse, this type of support work is a kind of open secret, ever present but rarely discussed. One newspaper article, for instance, featured the headline: "The Rush to Deploy Robots in China amid the Coronavirus Outbreak."[55] It spoke of staff shortages and "machine-dominated" assembly lines and seemed to suggest that full automation was just around the corner. Yet the image captions in the same article tell a more nuanced story of human involvement at every stage of this deployment. "A worker debugs a robot at a Sany Heavy Industry plant in Changsha" stated one. "Technicians adjust disinfection robots in a technological company in Qingdao" explained another. In these photographs, engineers stand around machines, often in pairs or groups, inputting instructions, tweaking movement, and grappling with problems. This is automation at the coal face, where theoretical assumptions must be adjusted for real-world conditions. Humans step in to adapt tech to local conditions, to provide course correction—or to rip off the panels and unjam an actuator when things go really awry.

Human labor also emerges when we peel back claims around "smart" or "real-time" warehouses. At Cainiao, the applications that support this array of functionality run in server containers, requiring intensive computing power. In a three-part series, engineers outline the work they have done to optimize these processes, employing an "elastic scheduling system," gathering metrics on existing use, establishing policies for allocating tasks, and building in toler-ance for glitches.[56] After deploying their changes, they monitor the impacts, correct for any mistakes, optimize if possible, and deploy again. Another post from the Cainiao CTO applauds how technical teams willingly gave up their holidays, "pulled all nighters" and "buried themselves in a lot more workload than usual" in response to pressures on technical systems during the pandemic.[57] While these details are highly technical, the point here is that the engineering team needed to carry out an enormous amount of meticulous work to sustain these systems. The performativity of the smart warehouse does not simply happen, but rather was constructed piece-by-piece through human labor and is constantly dependent on it for ongoing operations.

The same trends apply when we look to competitors like Amazon. On the Amazon Jobs website, there are currently over 7,000 postings with the title "Automation Engineer."[58] These positions range from Austin, Texas, to Newcastle in the U.K. and Hyderabad in India, leading to the inescapable conclusion that each fulfillment center requires a team of these engineers. According to one job posting, an engineer must "fully define performance on equipment, material and services," must be responsible for "incident follow-up, root cause analysis and documentation," must "support commissioning of new automation elements," and must "ensure procedural adherence to industry operating standards," along with a dozen other key tasks.[59] This work is highly skilled and highly paid. Crowd-sourced salary websites indicate that each of these engineers can expect to earn over $100,000 US per year.

Who trained these engineers, and what tools make their work possible? When we begin to map out the dependencies of human labor, it becomes clear that these engineers are only the tip of the iceberg. An engineer's expertise has to be built up over years, with her training drawing on the education sector and its extensive population of teachers, lecturers, assistants, and professional staff. And an engineer's contributions depend on the hardware and software sector—the chipmakers, developers, coders, and server technicians who build these systems and keep them running. "Even the most ingenious and accomplished automaton is far from allowing our hands to rest, much less replacing them," presciently noted F. G. Jünger more than seventy years ago, "for it is not a separate mechanism working by itself, but a part of a vast technical apparatus whose constant development entails an increase in the amount of work."[60] These observations highlight the army of labor needed for "automated" systems. They debunk the fiction of the lights-out factory and the human-less future of work that was supposed to be inevitable.

What, then, for the future? The move to digitization across various industries, coupled with the complex interactions of different systems and different standards, means that these kinds of support workers will be increasingly key. Implementation is far from perfunctory, and things will keep breaking. For sci-fi author Kim Stanley Robinson, it's clear that this labor of repairing and tending will not disappear anytime soon. In his book *New York 2140*, he describes the labor-intensive energy and transportation systems needed in a post-Anthropocene world, stressing the persistent role that humans will play. These systems "provided a lot of employment needed to install and maintain such big and various infrastructures. The idea that human labor was going to be rendered redundant began to be questioned: whose idea had that been anyway?"[61]

Such statements puncture the overinflated myth of full automation and reconfirm the centrality of human labor for the future. Certainly technology brings with it change. There is no question that work conditions in the twenty-first century will differ from the nineteenth or twentieth. And these shifts bring with them fundamental questions about the future of work. But this future is increasingly spoken of in terms of a collaboration between human labor and machinic processes—integrating rather than replacing workers. This integration may bring dignity, drawing on the worker's expertise and generously compensating her. Alternatively, this integration may be degrading, exploiting the worker and suppressing her rights. As this section showed, rose-tinted "integration" can quickly veer into feeling like instrumentalization pure and simple. Yet whatever the conditions, this more modest but realistic proposal prompts a sharper set of inquiries. What kinds of values and assumptions are embedded within these systems? Where exactly are the hand-off points in this mixed machinic-human labor? And what happens to oversight and ethics when automated decision making is employed? Here, retaining the fantasy of human-less labor in the face of reality becomes dangerous, obscuring the real issues occurring on the ground and the critical research needed to work through them. The next chapter delves into these new entanglements, a "future of work" in the present that is both strange and brutal.

2

Spotty Automation and Less-Than-Human Workers

"IT ALL STARTED SOME YEARS AGO," begins the narrator. "My hopes for a future in which machines would do all the work hadn't come true," he recounts. "Instead, I was working for a pizza delivery company." The voice describes how he masqueraded as a pizza bot, guiding customers through ordering and notifying them about their upcoming deliveries. In this story, a customer starts spamming the "bot" for fun, texting through puns and odes of affection. "We're sorry, we didn't understand—please confirm or decline," the narrator responds. But the customer refuses the instruction and continues his harassment. Finally, the narrator can't take it anymore. "Dude our automated system isn't set up yet," he blurts out, "This is a real person texting you. I make minimum wage. Please just tell me if you want the pizza!"

While this tale is fictional, taken from a work by media artist Sebastian Schmieg, it feels believable. Schmieg perfectly captures the deep entanglements between humans and machines that have come to define much of contemporary work. Automation promised to liberate us from labor, a future in which machines "would do all the work." Instead we are faced with systems that are semiautomated and devices that need augmentation. This is actually existing automation, a far more fragmented future of work that can rapidly shift from the promising to the pathological. Human labor is chopped up, split up, and farmed out, but never replaced. Humans are working harder than ever.

What does this work look and feel like? This chapter explores some of the new forms of labor introduced by "automated" systems. It zooms into workers operating at the coal face of automation, moving from digital piecework to machine minders and content moderation. These technologies shape work, allowing companies to offshore labor, lower wages, and restrict rights—in short, to more strongly dictate the conditions of labor. And these technologies shape workers, steering their practices and exerting intense social, cognitive, and psychological pressures on them. If automation is a myth, the necessity of propping it up nevertheless influences labor in profound and often brutal ways.

An Invisible Underclass

While the previous chapter touched on high-paid engineering work, automation introduces a spectrum of work, with a significant proportion being low-paid, monotonous, precarious—or all of the above. Work in this vein might include hand-labeling thousands of images for a data set that will be used for machine learning. It may involve moderating the unending stream of toxic comments appearing on a platform or flagging hundreds of videos with hate speech, animal cruelty, or illicit pornography. Or it could mean designing countless variations of an online banner, each with a slightly different message to enable microtargeting. This work is crucial. The smooth functioning of technical systems depends heavily on this kind of labor. And yet for political and financial reasons, technology companies often strive to maintain the illusion of autonomy. The "algorithm" makes those decisions; that's just how the "platform" works. Human labor is whisked behind the technical curtain, going on unnoticed in the background. Scholars Mary Gray and Siddharth Suri describe this hidden labor as "ghost work."[1]

In automation discourse, these jobs are often framed as a temporary state of affairs. Machines will soon oust their fleshy counterparts, we are told, but the technology isn't there yet. Homo sapiens is still needed for now, for those handful of tasks that haven't yet been totally routinized. In this framing, humans are temporary gap fillers that will quickly disappear when full automation arrives. And yet when we look to the real world, we see companies scrambling to deploy an ever-expanding range of technologies into new niches and new markets. Because there is no artificial general intelligence, each new context presents a slew of new problems and requires a specific solution. A technique to automate agriculture cannot be used in logistics. A model to read signs in one language will fail to read signs in others. A process that works flawlessly in the developing world breaks in the developed world. Each new issue means that the goal line of the soon-to-be-automated keeps getting pushed back. There are always new applications, new industries, and new contexts where computational methods are undeveloped and humans are needed. And even when a technical solution has finally been designed and deployed, the victory can only be described as partial. There are errors and oversights, features that need propping up with manual interventions. Together, these hurdles present what Gray and Suri term the "paradox of automation's last mile."[2] The grand dream of full automation will always remain unfinished.

In this context, humans become stopgap solutions, the "artificial artificial intelligence" needed for now. Technology companies often crowd-source this work, breaking up the labor into thousands of microtasks that may take a few seconds or a few minutes to carry out. These tasks are farmed out via a burgeoning array of crowd-sourced marketplaces, ranging from Amazon's well-known Mechanical Turk platform to more obscure players like Cloud-Factory, Playment, TaskCN, and Clickworker. Each task is simple for a human but difficult for computers. Is that object in the image a pedestrian or a power pole? Is that word a racist slur or harmless internet slang? Does that product belong in the kitchen or homeware category? Without this work, technical systems start to crumble in terms of functionality or user experience. Search features become useless; data sets are too messy or noisy to be valid training sources, news feeds fill up with spam or toxic comments. This is why AI and autonomous systems must "return to humans to backfill decision-making with their broad knowledge of the world."[3]

This work is low paid, with many microtasks yielding only a few cents. Workers generate a nominal wage by stringing together hundreds or even thousands of tasks per day. This situation has some appeal for those living in remote areas, who would otherwise struggle to find full-time employment. Its flexibility is also a draw card for those who need to juggle other part-time work or are caring for children. "As a woman, I prefer to work from home," stated one worker from Bangladesh. "I earn better than others. I have a child. I can maintain my family instead of doing a regular job."[4] And the ability for those in developing countries with a lower cost of living to tap into an additional income stream has proved to be compelling. Experts estimate there are now millions of active crowdworkers worldwide, and that number is only expected to grow larger in the future.[5]

Yet there is no question that this work is highly precarious. Scholars have challenged the claim that this work generates economic benefits, describing crowdworkers as a new "mobile underclass."[6] This is exploitative work, with little to no labor rights, conducted under intense time pressure. In this sense, this model is a form of "digital piecework," resurrecting the extractive and highly gendered home-based work, particularly in the garment industry, that labor activists worked for decades to abolish. Hidden away at home, on anonymous platforms, with no recourse when it comes to inequality or abuse, this new underclass largely goes unnoticed. The workers are still there but are intentionally concealed, suppressing their voices and ability to organize. As Veena Dubal stresses, digital piecework repeats a certain dynamic that we see again and again when it comes to capital and technology: "automation does not make labor obsolete; it reorders it, often rendering it invisible."[7]

Shitty Automation and Frontline Pressure

Automation, then, is incomplete. The previous chapter demonstrated how the dream of full automation never arrived. Instead, we are left with something partial and piecemeal. Workers are expected to join these pieces together. As Alexandra Mateescu and Madeleine Elish note in their study of retail automation, businesses "pick up and experiment with new technologies, relying on humans to smooth out the rough edges."[8] Their cognitive, social, and physical skills become a kind of bridging mechanism, connecting one partially automated process with another. People paper over the technical cracks. This work of sustaining and supporting technical systems is vital. Without it, "automation" cannot be successful. And yet according to the myth of automation, these jobs are obsolete, these roles on the way out. The fiction of full automation must tuck these messy human interventions out of the way. Mateescu and Elish observe that the "human infrastructure" that facilitates these systems is often invisible and undervalued.[9] Humans prop up technology while being overlooked and underpaid.

Instead of the smooth futurist imaginary that was promised, we are left with what tech journalist Brian Merchant calls "shitty automation."[10] Companies and institutions jump at the chance to cut costs, assured that these technologies will streamline their operations and erase their expensive labor pool. Yet after the hype of the promo video and the sales presentation, the actual systems are rolled out and intersect with real bodies and real spaces. Compared to the flawless vision, the reality is far messier and more mundane. Features only half work, parts break, and errors emerge. As Merchant observes, this is "half-baked automation" running on "ill-functioning machinery" with "garbage interfaces" that cause frustration for customers.[11]

Take the self checkout, for instance. By combining barcode readers, weight sensors, and software interfaces into a single machine, automation promised to delete that frustratingly expensive line item in the budgets of supermarkets and retail stores: the checkout operator herself. Leaping at this cost-cutting opportunity, companies have rolled out self-checkout systems across thousands of big box chains and grocery retailers. Yet rather than being erased, this labor is simply redistributed. One chunk is offloaded onto the customer herself. Now the customer must scan each item, ensure it has been correctly registered, and place everything one by one in the bagging area. While framed as a time-saving measure or a more convenient approach, this is ultimately free labor, a fact that doesn't go unnoticed. "I personally refuse to use the self check out because I don't work at the store," stated one customer. "And when I go to work, I get paid."[12]

Annoyed at this extra work, customers kick back against the system. Theft at self-checkouts is rampant. Incredibly, one survey suggested that up to a third of customers regularly steal when using self-checkout systems at supermarkets.[13] Techniques for exploiting these systems are gleefully shared among users on online forums. In the "banana trick," shoppers purchase expensive fruits or vegetables, but then enter a PLU number for a far cheaper produce item (usually bananas). In the "pay trick," customers tap the pay button, disabling the weighing sensor on some models, and then adding more items into their bags. And then of course there is the simplest trick of all: bag an item without scanning it.

If customers take on extra labor, so do employees. These systems were designed to do one narrow task and are totally incapable of handling the wider sociocultural problems they introduce. And so it is left to other humans to prevent theft, to help customers, and to work through errors and breakdowns. Workers' roles are now reconfigured. Now they become support staff, using emotional skills to understand why customers are becoming upset or aggravated. They become IT services, using technical skills to diagnose the cause of an error message or stoppage. And they become security personnel, maintaining line-of-sight across the machines and trying to spot the array of hacks that self-check swindlers use. As one sales manager stated, "if you're not paying attention and if somebody's not scanning an item, you're just giving it away, you know?"[14] All of these microroles come together to change the nature of the job. Supermarket employees must be sympathetic to customer complaints but also vigilant in spotting would-be thieves. They must be willing to stand on their feet for hours at a time but also tech-savvy enough to identify glitches and override defaults.

Far from erasing work, then, automation creates new work: new tasks and new problems requiring new skills. In this context, it is the frontline worker who is expected to step up and adjust herself and her work to these new sociotechnical conditions. It is the frontline worker who is placed under greater pressure, juggling tasks and augmenting so-called smart technology with actual cognitive and affective intelligence. And so in the end, it is the frontline worker, as Mateescu and Elish note, who must "absorb the frontline risks and consequences of cost cutting experiments."[15] Management invests massive amounts of trust and capital in automated technologies and then counts on employees to make it work. Harried and harassed, these laborers nevertheless press on, smoothing out the hurdles and humps of automation. In doing so, they turn a mishmash of sensors and systems into something cohesive—an integrated experience for consumers and a value-added product for their companies.

Machine Minding

Augmenting or hand-holding "automated" systems is not a new phenomenon. While the myth of automation promised that machines will soon improve and iron out all the glitches, the reality has always differed. Even as early as 1980, articles were mentioning the rise of "spotty automation," where machines had half-taken over the factory, carrying out the interesting tasks while leaving humans with menial ones. The result was a "corps of bored, dissatisfied people who do nothing more than act as machine minders."[16] That article, like so many others, suggested this was a temporary hurdle. Once companies embraced full automation, they assured readers, this awkward spottiness and the lowly roles it created would quickly vanish. Workers would soon be liberated to take up knowledge-based work, those high-skill, high-value positions like designer or engineer.

Yet rather than disappearing, machine-minding roles have persisted. "We're machine babysitters," state a trio of workers in Krish Raghav's graphic novel about these new labor forms in China.[17] "I take the queue numbers out from a printer and hand them over to guests," states one. "I jab a 'self-service' touchscreen on people's behalf," chimes in another. Why do they carry out such seemingly mundane tasks? "Our companies bought big shiny machines to 'improve service efficiency,'" explains one worker, "but they don't trust people to actually use them properly." These workers understand the hiccups and eccentricities of each machine. By navigating through menus, stocking them with paper, or even just turning them off and on again, they keep them operational and functional.

While Raghav's novel is fiction, it is based on his astute observations of new labor forms in contemporary China. And these machine-minding positions do crop up in real-world contexts across the globe. In England, one chilled-foods firm advertised for a "machine minder" at their facility. The minder would "take control of an automated line which moves the cooked product from the cooking vessel to different types of packaging, for example pots or pouches" and would be trained on how to carry out "basic machine maintenance."[18] In New Zealand, an agricultural company advertised for a machine minder who could "ensure the smooth operation of our automated packing equipment." The successful applicant would need to operate the machinery as well as "troubleshoot and resolve machine stoppage issues," "escalate problems as needed," and "monitor and replace consumables in the packaging, tubing, and lidding machinery."[19]

These roles arise because automation is incomplete. Placed under the pressure cooker of the market, companies have aimed to "modernize." The imperative is to move away from manual methods and toward routinized processes, to shift from hand-operated labor to standardized and mechanized production. Yet what often emerges is something much more mixed or motley, a worksite stranded midway between the fully automated and the non-automated. Processes require manual intervention to make decisions, deal with variability, carry out complex gestures, or even just to keep things moving smoothly. These workers, then, operate at the fuzzy nexus of partial automation. They are the human augmentation for machines that-are-not-quite-there-yet, bridging the gap between the breathless vision and the sometimes-broken reality.

Of course, this is not to claim that these are desirable positions. Spotty automation, as the `80s article suggested, often creates menial jobs, low-status roles that are rewarded only with low pay and few benefits. But this insight in itself is worth paying attention to, not least as a future flash point for labor relations. These jobs undermine the myth that automation replaces humans, a possibility variously framed as a labor-free heaven or an unemployed hell. Instead of these binary oppositions, we've arrived at something closer to a labor purgatory. Ours is a moment when machines work sometimes but not all the time, when some production but not all production is mechanized, when some use cases are supported but others must be manually assisted. Some things are automated, but many are not. For all the rest, automated technology must be supplemented and assisted, coaxed along by that most flexible and ingenious technology: the human body.

Backend Work, Hidden Trauma

If humans mind machines and step in for "automation" at brick-and-mortar stores, they also augment automation in more remote capacities. Here we turn to content moderation. Networked technologies have enabled anyone with an internet connection to post pictures, videos, audio, or articles online. These in turn can be viewed by almost anyone depending on their location and internet access. Content is the lifeblood of social media in particular. Without content, feeds grow stale, user bases dwindle, and advertisers leave for more popular alternatives. For this reason, social media offers tools to create, curate, and distribute content and offer an eager audience of consumers—"eyeballs" in the industry parlance.

And yet not all content is desirable or even lawful. Some of this material, such as child sexual abuse material, is illegal in many jurisdictions. Some of it, such as hardcore pornography, violates the license agreements or community guidelines of specific platforms. And some is simply deemed "objectionable" due to its language, propagation of disinformation, incitements to violence, or political viewpoint.

To filter out unwanted content, tech companies deploy a range of sophisticated algorithms that aim to identify and erase it. But while these automated solutions are beneficial to some degree, they are also partial. On an immediate level, material like memes, videos, and screenshots are often not easily machine readable. Understanding and extracting the "content" from a dimly lit scene, a busy protest, or a violent confrontation is far from trivial. Indeed, those who circulate such borderline content sometimes skew, reverse, or rotate imagery precisely to exploit these limitations. On a more fundamental level, the thin line between acceptable and unacceptable requires an understanding of context-specific norms and intentions. Navigating this highly ambiguous terrain, then, is still the remit of humans. "These platforms now function at a scale and under a set of expectations that increasingly demands automation," notes Tarleton Gillespie, but these "are precisely the kinds of decisions that should not be automated, and perhaps cannot be."[20] Is that image of a woman pornographic or meant to demonstrate breast feeding? Is that video from the Amazon a form of ancestral ritual, or does it veer into animal cruelty? These cases gesture to the layers of meaning that must be taken into account when deciding whether to keep or delete any piece of content.

Automated technologies fail, and humans must once again step in. Indeed, for content moderation providers, the known limitations of existing technical "solutions" are a clear selling point. In her book *Behind the Screen*, Sarah Roberts notes how some companies pride themselves on their manual approach, working it into their pitch. "We do human-content moderation," stated one firm; "we don't use robots, we don't use filters, we don't try to automate this process, we have human power, which, you know, it leads to a great deal of accuracy versus filters and tools."[21] The argument is simple but powerful: algorithms and automatic processes miss far too much. Vetting must be done by hand. Each piece of content, they promise, will pass before a human pair of eyes and be reflected on by a human brain. It's ironic, notes Robert, that instead of what was promised—"just wait until computers can do this better, that will unburden the human moderators"—we have the opposite: more people.[22]

In practice, this means an individual combing through articles, watching hour after hour of video, and inspecting thousands of posts. For their work *Dark Content*, artist duo Eva and Franco Mattes delved into this hidden labor, carrying out one hundred interviews with former content moderators.[23] Their stories provide a snapshot of the toxic fallout endured by those who undertake this work. One moderator described watching some kind of religious ceremony in which a young child was pierced with white-hot stakes and passed out from the pain. Another recounted witnessing a girl being brutally raped by two teens, with a third looking on and recording it all on his phone. Another said the most disturbing content he moderated was someone killing and dissecting cats by stabbing or electrocuting them, picking the "best scenes of the animals in pain or struggling," and then editing them together to construct a master montage of cruelty. "I can remove the content," stated one moderator, "but this shit is still in my head." After one session screening this material, he described "balling his eyes out" in his gray cubicle, heading to a bar after work to get extremely drunk, and then crying more at home.

These stories gesture to the psychological and mental damage that moderators absorb. These are the haz-chem workers of the digital world, who expose themselves to its radioactive waste day after day, week after week. The real issue "isn't the impact of individual images," stated one moderator who was contracted for Accenture for two years; "it's the accretion of thousands over time."[24] Exposure to this material takes a deep toll on the bodies and minds of moderators. While there are not yet long-term studies on this damage, research from other frontline roles like child protection[25] and law enforcement[26] demonstrates that workers experience high levels of stress, burnout, anxiety, and suicidal tendencies. Anecdotal evidence from content moderators sketches some of the same detrimental effects. Workers report feeling isolated and depressed, sometimes to the point of having suicidal thoughts.

One moderator, Selena Scola, claimed she developed posttraumatic stress disorder after nine months of viewing disturbing and graphic imagery on the job. She filed a suit against Facebook, which was joined by a number of other moderators. In 2020, the tech giant agreed to pay $52 million in compensation for mental health issues on the job.[27] Some of that money will go toward medical fees, with diagnosed workers able to access a licensed counselor and monthly group therapy sessions. The landmark ruling provides legal precedent for workers and organizations struggling to improve labor conditions. But the case also serves as a formal admission of the traumatic nature of this work. From drugs and alcohol to gallows humor, workers employ a number of coping mechanisms to deal with this damage. Yet their stories suggest these defenses are never enough. A toxic melange of cruelty and barbarity seeps into the worker's psyche over time, slowly injuring the inner life of the individual over weeks, months, or even years.

While Scola's win is commendable, it hinged on her being directly employed by Facebook. In one sense, then, the ruling only heightens the liabilities of "touching" any part of this labor process. Most moderators, in fact, work for third-party contractors. They may carry out work for high-profile tech titans such as Facebook, Microsoft, or Google, but they are officially employed by contract firms with names like TaskUs, UpWork, and Conectys. While these firms are headquartered in the US or the UK, much of the actual labor takes place in Global South countries such as India or the Philippines. And even when moderators work in-house, they are given different colored badges and barred from the same kinds of perks and privileges that "real employees" are given. No free lunches, no use of the recreational facilities, and most importantly, no included health care.

There is a kind of intentional distancing here, where these workers and their work are carefully bracketed off. For Roberts, this "feature" is a key part of how moderation has been set up and articulated.[28] Technology companies carry out a strategic "series of distancing moves" designed to "limit their responsibility for any workplace harm."[29] Even when the subliminal harms of moderating become overt, as in the case of suicide ideations or acts, companies can claim plausible deniability. It was the third party's protocols that failed. Their safeguards were insufficient. Their workplace practices were at fault. After all, moderators worked for them, not us. And these kinds of distancing tactics, as the next section shows, appear again and again when we look at automation and new forms of labor.

Extraction at Distance

Crowdsourcing. Gigging. Remoting. Zoom into many new forms of work, and you quickly discover that what is being automated is not the work itself but the delegation of that work. Networked technologies allow jobs to be broken out into tasks, tasks to be sliced into microtasks, and those tasks distributed to a pool of precarious laborers. "The human machine is there, pulsating and available," states Franco Berardi, and digital networks allow these cells of productive time "to be mobilized in punctual, casual and fragmentary forms."[30] A company's physical headquarters may be located in Beijing or London, but the firm is able to draw on workers living around the world, twenty-four hours a day, seven days a week. A mother in Bangalore wakes at 1:00 a.m. to feed her baby and then jumps on to do a twenty-minute stint. A student in Manila churns through several tasks during his lunch break.

What is truly new about software, argued Tessa Morris-Suzuki in a remarkably prescient essay written almost forty years ago, is that "the worker's knowledge may be separated from the physical body of the worker and may itself become a commodity."[31] Traditionally, the site of labor struggle has been the warehouse, the factory floor, or even the office. In this conventional model, workers travel to these sites and create value by employing their body and mind. Equally, they may choose to withdraw their laboring body from these sites in the case of strikes or walkouts. Yet network technologies and platform labor alter this long-standing paradigm. Laborers and the means of production are decoupled. Humans and machines are split apart. "We are left with, on the one hand, machines which work automatically, endlessly responding to the instructions provided by workers who may be physically far removed from the production site," notes Morris-Suzuki, "and, on the other, the increasing channelling of living labour into the process of designing, composing and altering those instructions themselves."[32] Once remote workers are signed up and signed on, their labor can be captured in any number of ways. That tiny piece of intuition can be stored; that small gesture can be recorded; that emotional reaction can be snapshotted. Farmed out and then gathered up, all of this labor is funneled back into various technical systems, improving models and enhancing user experiences.

Technical systems glom onto the human, extracting its useful qualities. As my first book argued, the imperative here is not to "employ" labor—which would require sustaining it with a living wage, comprehensive health care, employee benefits, and so on—but to extract it while retaining legal, financial, and geographical distance.[33] The aim is to scale up and out, expanding operations while insisting on the legal and ethical autonomy of workers. Given this drive, it is no surprise to witness the rise of companies like Uber and Airbnb premised precisely on this model. "Uber, the world's largest taxi company, owns no vehicles," proclaimed tech pundit Tom Goodwin, "and Airbnb, the world's largest accommodation provider, owns no real estate."[34]

In this model, workers are no longer employees but freelancers defined by neologisms like Uber's "driver-partner." These workers must manage themselves, insure themselves, and promote themselves. They must constantly adapt their practices to the app's algorithmic management, which can shift rapidly from one day to the next. And they must carry out their tasks consistently and professionally, with the pressure of bad ratings and permanent bans constantly hanging over them. For the company, this is a flexible labor pool that can stretch and flex on demand. Yet just as crucially, this is a decoupled labor pool, cleanly disconnected from the rights that usually come with employee status. The company remains asset-light and agile; the worker takes on the heavy externalities and anxieties needed to earn a living.

Automated technologies, then, assist in draining labor from afar. They allow companies to exhaust the productivities of workers without being saddled with liabilities. Marx was astute in recognizing that this extractive relationship was at the core of the value relation. "Capital is dead labour, that, vampire-like, only lives by sucking living labour, and lives the more, the more labour it sucks," he famously observed.[35] Similarly, Luciana Parisi has described this as a "human-machine mode of parasitism."[36] Crowdsourced or platform-mediated labor, particularly in our post pandemic period, aims to extend this parasitic mode to more workers in more remote locations.

Rather than replacing the human, Parisi points out, such crowdsourcing platforms facilitate "less-than-human" labor, exploiting workers to carry out machine-like service.[37] How is labor made "less-than-human"? Through that most ancient of technologies—that difference-making device we refer to as race and gender. The lack of an appropriate algorithm, observes Parisi, means that high data tasks must draw on a "racialized and gendered pool of global temporary workers."[38] Following the logic of capital, the aim is to keep identifying ever more precarious labor pools in order to keep pushing labor costs lower. According to this brutal calculus, the remuneration and conditions should be reduced until they reach an absolute baseline beyond which people will not engage in work. And who sets this baseline? By operating a global marketplace, this baseline is established by the most marginalized individuals who can somehow gain access to a computer or mobile device and an internet connection. A number of crowdworking platforms operate "Cheapest Offer" auctions, where workers compete against each other to offer the lowest possible price. Scholars have shown how this logic creates "wage dumping,"[39] plunging the price downward to the point where only the workers with the lowest cost of living can generate income.

Like a homing beacon, then, these platforms are hunting for surplus populations composed of the unemployed or underemployed. And these populations are invariably people of color, or women, or those in the Global South. As Niels van Doorn observes, "the majority of cleaners, janitors, and home care providers operating in the gig economy are working-class men and women of color."[40] Yet the real "triumph" of digital platforms is to use the labor conditions for the surplus population as the template for the whole population. Their "innovation" is to make work, regardless of who carries it out, racialized and gendered. There are certainly white men operating on these platforms, driving cars, running errands, or completing microtasks while referring to this labor as their "side hustle." However, as van Doorn stresses, "they are unable to fully detach from the gendered and racialized order of worth that haunts this kind of service work and continues to associate it with a lack of value, skill, and dignity."[41] As the conditions and paycheck quickly reveal, they are laboring in roles that have leveraged racialized and gendered stereotypes to render this work low status.

All the Work without the Workers

Alex Rivera's *Sleep Dealer* anticipated this future of labor.[42] The science fiction film is set in a near-future scenario in which a massive wall permanently separates the United States from Mexico. Rather than crossing physically into the country, Mexican emigrants work from digital factories. By hooking their bodies up to these networks, they "plugin to the new American dream," remotely puppeteering robots that carry out labor in the States. These robotic proxies perform a broad range of work, from building skyscrapers in Chicago to picking oranges in Florida and packing meat in Iowa. The grand dream of "automating" the full spectrum of work has finally arrived, and yet it comes not through some groundbreaking advancement in general intelligence, but by a Mexican performing to the point of exhaustion. Released over a decade ago, the film is uncanny in its prediction of a number of labor trends. Yet for Rivera, this vision simply extrapolates from the core logics that have always dictated labor relations. "Essentially, our economy has no moral core or moral guard rails on it," he notes, "If you know that, and look at new technologies, you can quickly arrive at the predictions I made."[43]

The film undermines the myth of automation, asserting that these technologies are not about erasing work but about displacing it. Labor is shifted onto women, migrants, people of color—subjects typically less able to exert their rights and make demands. Technology's "innovation" is drawing on these laboring bodies while keeping them far away from influential clients, from corporate headquarters, and even from other laborers they might collectively organize with. Companies want to harness these skills and abilities while ensuring their political agency remains at a safe distance. Digital networks allow them to push this problematic body far enough away so that it, along with any claim-making potential, effectively disappears. The dream, as *Sleep Dealer*'s trailer proclaims, is simple but brutal: "All the work, without the workers."

In the end, this is how automation "erases" the human. The full human subject—that rich historic figure endowed with inalienable rights who can draw on a potent spectrum of claims—is deleted. Pasted in its place is a more shadowy figure who is "less-than-human" by design. This is the global digital worker, who may appear only briefly as an avatar or a username. For this figure, agency has been ring-fenced, shrunk to an ideal zero point. Conditions are locked in place by the platform and its set menu of features and functions. The pay rate is fixed and the choices are few: to work on this task, to work on that task—or not to work at all. They are "free" to provide their labor and free to withdraw it. If they provide it, they must compete with others across the globe for the same tasks, carrying them out faster or cheaper. Technology's triumph, then, is turning all of these laborers into racialized and gendered laborers whose working conditions are defined by a race to the bottom. The global digital worker is a figure whose work is so simple that "anyone" can do it and whose identity is so generic that "anyone" can replace them.

Full automation, then, never arrived. Total autonomy was never achieved. Instead we are left with a future of work that is half-baked, an array of semiautomated systems and processes that need augmentation and intervention. And yet in one sense, the dream to replace the human laborer came true. The human worker, on full pay, with full hours, and full benefits, is becoming harder to find in many contemporary labor regimes. For companies that have embraced new schemes like platform labor, crowd-working, and gigging, that person is now a relic, a vestige of a former time that is unlikely to return. Instead we see the rise of a more anonymous and precarious laborer, a lower-paid and lower-status individual who can be drawn on when needed and dispensed when necessary. Now, "just-in-time production is overseen by a permanently temporary labor force."[44]

Undermining the myth of automation and developing a portrait of actually existing work points to this figure and the conditions she grapples with. This insight brings with it a certain bleakness. Faced with falling wages, reduced rights, and shrinking agency, it is easy to slide into fatalism. To be sure, suggesting that labor schemes revert to the halcyon days of the past seems like wishful thinking. And certainly to declare that any single intervention—from government regulation to reeducation schemes—is the silver bullet solution for all workers everywhere would be disingenuous. What the response should be is unclear. Yet new conditions always usher in new capabilities. Digital technologies unlock new forms of organizing and mobilizing. Platform mechanisms used for exploitation can be co-opted for solidarity. And the fact that automation is incomplete means that human labor is valuable and irreplaceable. All of these factors should offer hope to those seeking to build a more radical and sustainable future of work.

PART 2

The Myth of Automation Everywhere

3

Technology in Context, Technology as Culture

AUTOMATION IS A MYTH because automation is universal. Whether described in boosterish business terms or predictions of doom, automation is framed as a ubiquitous force rippling across the globe. Best sellers warn that an "automation wave" is about to hit the world.[1] Business reports echo this understanding, describing three "waves" that will sweep over society.[2] For other pundits, automation is better described as an "age," a new era that will usher in global change.[3] The same all-encompassing rhetoric can often be witnessed in discussions of automation, which quickly leap into a highly abstracted debate employing grand terms like "humanity," "technology," and "society." Together, these terms work to toss out the specificities of city and country, culture and region. Automation, it seems, is a universal condition, a de-situated force that will soon envelop the planet.

Take *The End of Work,* for example, Jeremy Rifkin's widely sold book that lays out the argument in its title. "We are entering a new phase in world history," intones Rifkin, "one in which fewer and fewer workers will be needed to produce the goods and services for the global population."[4] What will be the cause of this dramatic new epoch? "More sophisticated software technologies are going to bring civilization ever closer to a near-workerless world," he explains. And these technological advancements are poised to impact every country and every industry. All sectors of the economy, warns Rifkin, are now "experiencing technological displacement, forcing millions onto the unemployment rolls." No matter where you are or what you do, you will soon feel the ubiquitous force of automation in the form of wall-to-wall redundancies.

Automation, then, is framed as a global phenomenon, an overarching force that floats above the particularities of this culture or that culture, this place or that place. This is a "world without work,"[5] a "time of brilliant technologies"[6] that will soon envelop us all. Such framings flatten a vast constellation of distinctions, erasing differences in language, history, socioeconomic status, labor rights, political regimes, and numerous others. The resulting myth is powerful but problematic, collapsing cultural difference to tell a unified tale about the future of technology and its impact on labor. Certainly new technologies usher in shifts—changes in labor conditions that matter deeply for workers, employers, governments, and society in general. And yet in refusing to be pinned down, automation loses much of its analytical purchase.

Where exactly is automation to be found? In many popular books, the unit of analysis is the world—"where" effectively becomes "wherever." The result is that the same fate seems to await everyone, everywhere. The use of the singular betrays this shared destiny. In their first book, Brynjolfsson and McAfee want to tell you how the digital revolution is transforming "the economy." In their second, they aim to sketch out the influence of robotics on "the labor market." And in his best seller, Ford discusses the jobs and threats of unemployment in "the future." But as any geographer or ethnographer could tell you, there are economies plural, markets plural, and futures plural. This basic fact suggests that technology will be adopted differently in different places. And these distinct approaches and relationships, occurring within the local lifeworld of individuals and communities, will produce a set of highly divergent futures.

Technology in Context, Technology as Culture

So what—and where—are we talking about when we talk about automation? Dig into these systems and you quickly discover their sitedness, their grounding in a particular time and place. Instead of accepting the 10,000-foot view offered by automation discourse, it pays to focus on particular instances of automation. Zoom in, and you might arrive at a warehouse in

Hangzhou, for instance. Dotted around the floor are a swarm of automated guided vehicles, the robots that shuttle crates back and forth. Pick one of these devices up and unscrew the cover, and you'll find a processor chip not made by U.S. semiconductor giants like Intel or Qualcomm but by HiSilicon, a Chinese manufacturer. This tiny chip points to an important tale.

Several years ago, the inroads that Chinese firms had recently made in the fields of machine learning, wireless protocols, and particularly in 5G—a key critical infrastructure technology for the foreseeable future—caused unease in Washington.[7] While these corporations were multinational, the Trump administration believed they maintained close ties with Beijing and that this undue influence would lead to backdoors or other surveillance mechanisms being embedded into future information technologies.[8] They responded by blacklisting information technology providers like Huawei and Xiaomei, along with dozens of other Chinese tech firms. U.S. companies were prohibited from purchasing parts from these companies or supplying electronic components to them.

These moves escalated into a "technology war" that meant that cooperation or even communication between Chinese and American tech firms was prohibited.[9] Chinese companies responded by starting to domesticate their supply chain. Of course, global dependencies are complex; this process will not happen overnight or perhaps ever in totality. Yet one by one, American components have begun to be stripped out and alternative Sino suppliers found. Technology wars and the regulatory red tape that accompany them are not going away, forcing international alliances to be substituted with technology that is designed and developed locally.

These shifts exemplify a turn toward nationalized infrastructure. "In China," notes Gabriel de Seta, "two decades of state-led ICT development and a conception of cybersovereignty elevated to foreign policy spearhead have carved out a geopolitical enclave in which computational architectures and informational actors are coming together into what could be deservedly termed the Red Stack."[10] This stack is at once political and technical, a homegrown assemblage that starts at the bare metal of the chip and moves all the way up through operating systems and wireless protocols to the software applications at the top. For states, building out the stack is a way to wrest back control over the digital technologies that increasingly shape everyday life, developing an end-to-end ecosystem with local and national principles at their heart. This is "our" tech with "our" values.

The domestication of technology points to a basic but fundamental point: automation is not a universal state but rather a fragmented condition. Looking across time and space, we see a patchwork of wildly differing approaches to simulating intelligence, optimizing work processes, and carrying out human-like gestures with mechanical devices. Some technologies were funded by the aristocracy; others were commissioned by the industry or the military. Some devices were designed to attack strongholds or wound enemies; others were constructed to keep accurate time or entertain the court with song and dance. All of these projects were embedded in a distinct cultural setting, from upholding certain notions of power and prestige to practical aspects like funding, design, and fabrication. Far from being a single red thread, the development and adoption of technology is more like a rich tapestry composed of powerful patrons, local inventors, specific communities, language adaptations, regional needs, and many other interwoven elements.

This kaleidoscope of particular people and places erodes the universalism of the automation myth. For Yuk Hui, we need to reject the universal history of man and the universal concept of technology. "What we lack now," he stresses, "is the thinking of a technodiversity in the current geopolitics."[11] Hui reminds us that technology is always deeply situated, grounded in a particular place and time. Context shapes how these techniques, infrastructures, and processes will be taken up and implemented. Just as there is no singular television but rather "this television, our television," there is no singular technology; technologies are plural and local, and they must be understood as a "cultural form."[12] Hui makes this same point by excavating a distinctly Chinese philosophy of technology from traditional Daoist and Confucianist thought, arguing that there is not a single technics but multiple technics.[13]

Building on these insights, we can assert that there is not a single "Automation," but rather many automations. Each of these versions is distinct, varying from place to place. Automation in China unfolds differently from its deployment in the United States. Automation in India is following a distinct path from its development and adoption in Australia. This is not to descend to a kind of cultural essentialism, where tech gets read through broad stereotypes, but to simply acknowledge that technology and culture are deeply intertwined—so entangled, in fact, that it is often impossible to distinguish among them. These formations emerge from the unique interplay of cultural values, local expertise, research institutes, supplier ecosystems, and state intervention in each nation. Technology, as Vincent Mosco reminds us, is "mutually constituted out of a culture that creates meaning and a political economy that empowers it."[14]

Stolen Tech and Shanzhai

Without this cultural perspective, misunderstanding is rife. Automation is sometimes discussed in a broader debate around AI technologies, chip production, and a long-standing Sino/American rivalry. Here, Western commentators have been acerbic, bitterly throwing out accusations of "stolen tech."[15] Chinese companies have blatantly ripped off their U.S. counterparts, they allege, imitating hardware designs and software implementations. Rather than beginning from a blank slate and forging their own path, they have taken a shortcut, borrowing and adapting from others. Blind to their own cultural bias, these Western pundits assert that creation-from-scratch is key, that originality is everything. In refusing to play by these rules, Eastern competitors have cheated, committing the mortal sin of copying.

But as Byung-Chul Han asserts, a "prior, primordial positing is alien to Far Eastern culture."[16] In this context, the concept of an original, a unique object without precedent, is out of place. His book *Shanzhai* explores the Chinese cultural trope of the same name, a term that literally means "mountain fortress" but now refers to counterfeits or knockoffs, particularly in the electronics sector. Han goes beyond these Western slurs to excavate the history of this concept, tracing the fundamentally different way that imitation has been viewed within that culture. Here, replication is a sign of success, with the most desired objects being copied the most. And here, reproduction is a kind of homage, with each successive variant paying respect to earlier traits while also striving to improve upon it.

In the West, "shanzhai" has gradually moved from a term of derision to a concept that draws begrudging praise. Shanzhai's approach of creating countless product variants for underserved niches has led to some wildly creative designs, from phones with built-in Buddha shrines and cigarette holders to watches with horoscope maps. For those in technology sectors and business schools, shanzhai injects something new and enticing into tired capitalist enterprise, constructing a novel innovation ecosystem that clearly feeds into long-tail market demand while retaining a number of uniquely Chinese characteristics. And yet, as Silvia Lindtner, Anna Greenspan, and David Li note, shanzhai is "neither straightforward counterculture nor pro-system."[17] It is certainly entrepreneurial, focusing closely on market dynamics—but often plays on the margins of those markets. It is hard-headed and profit aware—but its designs can also be seen as social projects in recognizing religious devotees, the elderly, and other communities typically overlooked by global brands. These qualities create a unique form of cultural production that is "critical enough to resist assimilation yet constructive enough to have a sustainable impact."[18]

Shanzhai thus walks a kind of tightrope, maintaining a delicate balance that acknowledges the power of market forces while stubbornly holding onto a degree of creative independence. When it comes to automation, then, we should expect that shanzhai culture would retain this ambivalence, feeding into local technical initiatives but also tweaking them, twisting them, or even working against their worst excesses. In their fieldwork from Chinese maker spaces, Xin Gu and Pip Shea observed that local communities resisted a "universal maker culture" and saw "Shanzhai 2.0 as more appropriate in addressing local problems such as automation and gentrification."[19] In this sense, shanzhai is both the automated assembly line and the home-grown phone or hacked-together wearable that helps workers cope with the pressures of production.

Shanzhai, then, sheds light on local approaches to technology while refusing to ever be pinned down entirely. This disclaimer can be applied to cultural insights in general. Along with differences among cultures, there are also major differences *within* cultures. While companies like Alibaba and Didi Chuxing emerge from China, it's clear they adopt a range of different philosophies. Some founders embrace Confucian principles such as "harmonious while different" to guide their decisions; others cherry-pick a mix of practical strategies drawn from the East and West. Yet at the very least, bringing the cultural context into this conversation can provide a deeper understanding of a firm's strategies and motivations: where they choose to concentrate resources, how their research is carried out, the path that their product development takes, and so on. Decisions around automation and technology more broadly are not purely utilitarian but are shaped by the culture they are embedded within.

Ethics and (Dis)trust

Folding in the cultural dimension may also provide insights when it comes to the ethics of technology. The rise of automation and AI has seen a cottage industry emerge around these topics in the West. A wide array of institutions, task forces, and organizations have sprung up, each offering its own moral doctrine for this brave new world. From the "pillars of ethical automation," to a four-stage model for "socially responsible automation" and the "ethical auto-mation toolkit," a dizzying number of frameworks, guidelines, and principles are on offer.[20] Yet as Han argues, "Chinese thought distrusts fixed, invariable essences or principles." For Han, the Chinese character *quan* embodies this fluid approach to thought, denoting a kind of weight that can slide back and forth on a scale. "Quan describes the potential inherent in a situation rather than a set of rules that remains the same, independent of conditions and situation," he explains. "In the context of quan, nothing is final."[21]

This more fluid conception of ethics is reminiscent of machine learning, providing a set of parameters or nodes that can be tweaked to reach a desired outcome. Rather than a rigid moral code, quan is more like a dial that can be ramped up and down. Of course, it would be too simplistic to say that quan or any one trope feeds directly into a definitive Sino technical ethics. These concepts instead constitute a kind of cultural grammar that institutes and actors can draw on when creating ethical frameworks for emerging technologies. Both China's Ministry of Science and Technology and the Beijing Academy of Artificial Intelligence have published their own sets of principles in recent years. These guidelines, together with specific examples of technology in the field, start to mark out a distinctly Chinese approach to digital ethics. While Western frameworks tend to focus heavily on the rights of individuals, scholars anticipate this ethical approach will emphasize social responsibility and group relations.[22]

It's clear that culture also shapes the adoption of automated technologies for end users. Over the past two decades, a string of studies have investigated why certain cultures tend to trust automation more than others. One study compared how users from the United States, Taiwan, and Turkey interacted with an automated path planner.[23] The scholars were interested in whether the "cultural logics"[24] model might explain how these technologies are approached. This widely used model defines three cultural types: Dignity, Face, and Honor. In Dignity cultures, prevalent in Western Europe and North America, self-worth is derived internally. In Face cultures, characteristic of East Asian societies, self-worth is derived externally from a stable social hierarchy and fulfilling role obligations. And in Honor cultures, associated with the Middle East, Latin America, and the Mediterranean, self-worth is linked to honor, which must be claimed as well as paid to others.

As the researchers predicted, cultural background deeply shaped the way that the same technology was treated by each group. The Turkish group displayed the "slow-trust" characteristic of Honor cultures, with this distrust ultimately leading to disuse of the technology altogether. The Taiwanese group, in contrast, tended to display automation bias, relying on its decisions "even under conditions of reduced trust." Based on these distinct responses, the researchers recommended that multiple interfaces be developed, surfacing more information for some cultures and giving others more control. Granted, any typology is prone to generalizations. Cultures are certainly not monolithic, and individuals vary in their embrace of traditional cultural characteristics. But these insights add in a much needed cultural dimension to the universal framing of automation that is typically put forth. Rather than a bland globalism, they provide taste and color, a specificity that is sorely lacking from most analyses.

Automation or Consolidation?

One benefit of a more situated approach to automation is that it provides conceptual clarity. Rather than ascribing an outcome merely to "the technology," a socioculturally attuned view highlights the local dynamics at work.

A tech company in Shenzhen, for instance—a location known for its vast hardware ecosystem and shanzhai-inspired approach—may take a completely different approach when it comes to technical development. There is a sense of "thinking-from-within" here, of the researcher, policymaker, or technologist placing themselves in another company or organization's shoes. From this perspective, a whole new set of forces comes into view: the history and norms of an industry, governments' laws and incentives, regional strengths and weaknesses. Of course, that doesn't mean technical aspects are irrelevant. Information infrastructures and technical functionality clearly feed into any discussion around contemporary labor regimes. But in some cases, technology seems to be the least important part of the story.

Take the origin story of Cainiao, for example, the Chinese logistics giant. In 2010, the Chinese logistics market was fragmented and inefficient. The status quo was for companies to contract separate firms for logistics, storage, and warehousing and turn to independent service providers when it came to supply chain management. This piecemeal approach meant that companies might have a dozen firms handling different aspects of their business. Much of this fragmentation was a result of local protectionism. "Driven by the desire to maximize local economic growth, employment and tax revenues, and less by concern about the efficient utilization of regional resources," local governments had lobbied for transport infrastructure funds even when they weren't necessary.[25] This created duplicate logistics parks, a glut of providers, and general oversupply. Firms proliferated to the point where the top twenty logistics companies in China controlled only 5.5 percent of the market. This chaotic situation was made worse by the lack of standards when it came to practice and procedures. Factions of small companies engaged in price wars and depended heavily on handwritten bills of lading. This was the "logistics dark age."[26] The result was predictable: a market "saddled with poor service standards, ineffective package tracking methods, and dismal delivery times."[27]

Cainiao entered the market in May 2013, aiming to rework every segment of the product chain to draw on the benefits of contemporary technology. The concept in essence was relatively straightforward: Cainiao would build a platform for intelligent logistics and handle the data while collaborating with leading logistics providers to handle the warehouses, vehicles, and people.

The original consortium included companies such as Yintai Group, Fosun Group, and Forchn alongside delivery companies like Yunda and SF Express. Yintai's specialty was warehouse management; Fosun focused on warehouse construction; Forchn was an expert in line hauling.[28] This approach established an automated end-to-end supply chain.

This story, while itself bordering on business creation myth, stresses consolidation rather than merely "automation." It is not that Cainiao simply applied technology to various segments of the logistics supply chain, but that it was able to integrate them, to bring them under a single banner. Rather than contracting a dozen companies to manage storage and shipping, one provider could coordinate it all. We can see this consolidation at work simply by looking at the logistics pieces that Cainiao has assembled: machine learning techniques, standardized logistics databases, last-mile delivery, and local collection points, among others. Technically, of course, many companies are involved in this chain, from the small businesses making the products to the Tmall storefront that sells them and the various last-mile contractors that deliver them. This is what scholars call an "interorganizational network,"[29] where strong links between companies boost their reach and performance far beyond what they could accomplish on their own. In this sense, Cainiao is much closer to an ecosystem model than a single corporation such as Amazon.

Yet whether framed as ecosystem or platform, Cainiao's consolidation is powerful in establishing certain standards and procedures that companies must adhere to. For instance, Chinese addresses often vary significantly, with many different formats given. Some place the postal code before the province, some after it. Older or less modernized styles group the district and the city together. To combat this variation, Cainiao makes databases of standardized addresses freely available to a range of partners and companies. These four-level addresses contain "standardized and structured administrative address information," establishing a common way to designate a location.[30] This reduces duplicate databases that may be incompatible with this system and has flow-on effects in terms of better dispatching, improved path planning, and optimized vehicle routing. Standardization, then, is more than merely aggregation: this is not just about bringing all this information together under one roof. For data to be useful, it must be compatible. Users must be able to cross-index fields and compare values against each other. As Ned Rossiter reminds us, "logistical technologies derive their power to govern as a result of standardization across industry sectors coupled with algorithmic architectures designed to orchestrate protocological equivalence and thus connection between software applications and workplace routines."[31] Entities and agents across multiple sites must be made consistent so that they can be identified, evaluated, and optimized.

By imposing a unified schema across the platform, Cainiao establishes a degree of interoperability. Processes that were seen as distinct can now be related; activities in one part of the supply chain can now feed usefully into another. Yes, Cainiao may have beaten other companies in any single segment, tracking packages or shipping faster. But consolidation stresses the new connections that have been forged across a local context and a particular supply chain. This new cohesion allows the links, the intersection points between things, the transitions across phases to be considered. Here we're reminded of Sandro Mezzadra and Brett Neilson's focus on the production of operations "not as a 'thing' but rather a set of links or relations between things, which is to say the framework or skeleton of a world."[32] This insight pushes back against automation's mantra of better, faster, more efficient. Cainiao's story is less about technology "making things better" and more about putting people and things into tighter relation with each other. Processes that were disjointed are now interoperable.

Consolidation forces objects and events to be compatible. Now data generated by one activity can be used at another segment of the chain. The Double 11 festival, for instance, is the largest shopping holiday in China. From both a retail and logistics standpoint, it is hard to convey the sheer magnitude of this event, with sales dwarfing Black Friday and Cyber Monday. In 2020, gross merchandise volume was $74 billion U.S. dollars, with Cainiao processing 2.32 billion orders over the eleven-day period.[33] Rather than simply guessing what consumers will buy at the next festival, Cainiao does extensive data mining on historical numbers to generate predictions. For businesses, these forecasts guide stocking decisions, reducing their prep time from months to weeks. For logistics companies, the forecasts help to prepare fleets and routes, reducing delivery times to the next day or even the same day. In the future, the same type of purchasing data could also be used on a highly localized level. Cainiao plans to launch 30,000 "community post" stations across 300 cities in China. Based on its wealth of purchasing data, the firm could identify the most popular products in each locale and stock them at these post stations.[34] The result for consumers would be a form of instant delivery.

Of course, the relationships key to consolidation are not always smooth. As companies increasingly recognize the value of data, access and ownership can easily become a source of conflict between partners. Data becomes a wedge rather than a bridge. One clear example of this was the spat between Cainiao and SF Express. SF Express is one of the key courier services that Cainiao used to provide last-mile service, delivering around a million packages per day. But in June 2017, it suddenly stopped sharing its data with Cainiao. Blocked from this tracking information, merchants had no idea if or when their packages had been collected—and when they could expect to be paid. SF Express was defiant, stating it would "continue to guard its core competitiveness and hopes other courier companies will be vigilant against Cainiao dipping its finger in courier companies' core data."[35] Cainiao responded with force, immediately suspending SF Express from its network. The event triggered a dive in the courier's share price, rapidly wiping 8 billion yuan off its value.[36]

While this particular dispute has now been resolved, we can expect that clashes over data will only grow. These struggles are a reminder that the data so crucial for automated technologies is not a static entity that exists in a vacuum. Data must be constantly produced, collected, and shared on a daily basis—and this ongoing performance depends on mutual agreements among parties. These agreements are cultural in that they draw on societal norms and shared understandings. And these agreements are political in conferring certain advantages, shaping rights, and structuring power relationships. Data foregrounds the local context that swirls around any given automation technology, a whole cast of corporations vying for market dominance, organizations jostling for recognition, and individuals striving to assert their rights. This specific milieu reminds us that automation is occurring not just anywhere but somewhere.

Domesticating Automation

To speak of a concrete "where" is to push against a vague "wherever," the universalizing tendency of automation discourse. As we saw in the introduction to this chapter, these texts are dominated by a global perspective, suggesting that whatever technology has in store for us, it will happen to everyone, everywhere. But if we can learn anything from adjacent histories of technology like the internet, it is that nations are increasingly dissatisfied with the "global." "Global," in fact, frequently means U.S. dominated in practice, and part of this pushback is a rejection of the American values embedded in everyday technologies. Critics have long observed that tech companies like Facebook, Amazon, Google, and Apple all emerge from a "California ideology," a "dotcom neoliberalism" that seamlessly blended the spirit of the hippies with the market mind-set of the yuppies.[37] Sprung from this particular cultural context, their products and services are more or less aligned with the technolibertarian ideologies of Silicon Valley and broader Western axioms like consumerism and individualism. These values are embedded in the platforms that millions use daily, a situation that some find intolerable. These "digital dictatorships transcend traditional national borders, enforcing their beliefs, narratives and rules on the world at large," stresses Kalev Leetaru, effectively creating a new form of "cultural colonialism."[38] For states with more authoritarian leanings, a shift from the global internet to a national internet is crucial, allowing them to start stripping out unwanted values and begin embedding their own ideals.

In the past few years, the same domestication can be witnessed in artificial intelligence. AI is increasingly seen as a "foundational technology that can boost competitiveness, increase productivity, protect national security, and help solve societal challenges," and so nations are not content to let the United States dictate this space.[39] Certainly economic concerns are a key driver here. AI is seen as a key to unlock new sources of value and new markets, bringing financial prosperity to the country able to achieve a commanding lead. But cultural factors are increasingly on nations' radars as well. "The United States needs a human-centered AI framework that guides a national policy and plan," stressed highly influential computer scientist Fei-Fei Li, "a framework anchored by our shared American values of equality, opportunity and agency."[40] Early movers get to put their stamp on tech, embedding their ideals and ideologies at the heart of critical information infrastructures. As a result, China, the United States, the European Union, and a dozen other countries are all engaged in an intense AI race. Through strategic combinations of policies and programs, each seeks to accelerate research, attract talent, and claim the next major breakthrough. Rather than promoting collaboration among allies, countries are increasingly digging in, choosing to develop tech products and programs that are consciously homegrown. States are moving from internationalism to the indigenous.

When it comes to automation, then, we see the same sort of nationalization at work, some of it intentional but much simply a by-product of any local context. First, each country has its own unique blend of software developers, hardware suppliers, research and development laboratories, and academic institutions. This domestic ecosystem has certain strengths and weaknesses. A confluence in factors in China, for example, has meant that it excels at facial recognition intelligence and applications, while lagging behind in general AI research and chip production.[41] These characteristics in turn privilege certain aspects while downplaying others. Chinese companies, according to one industry insider, place a premium on applying technology in practical ways to generate profit while accepting imitation as part of the landscape.[42] These specificities explain the power enjoyed locally by many platforms—but also some of their difficulty in spreading behind the Mainland. This domestic ecosystem underscores the basic but important point: technologies will be conceived, developed, and adopted differently in different locations.

Second, the economic and policy environment differs enormously from place to place. Some countries place immense amounts of resources and faith in automated technologies, while others view them as an expensive failure, a string of broken promises. As one article quipped, "automation" is a buzzword for some governments and a dirty word for others.[43] This view determines the particular mix of political and financial incentives that are offered around technology development. Enticements might include a laissez-faire approach to regulation alongside long-term loans and tax breaks. One Shenzhen firm focused on automation disclosed that it was receiving large payouts from the city of Shenzhen as well as a state grant to produce a new factory, where "the land, the factory and even the office furniture are all offered by the government for free."[44] These kinds of measures lower the risk for companies, taking the sting out of the massive capital outlay that is often required for the design, tooling, and purchase of such technologies. Of course, expectations may not match reality, as some companies have found. After having a cobot for a year, "we realized that it did not make any sense," admitted one firm; "we took it down and sold it."[45] Yet for better or worse, this policy environment shapes the local technical environment, influencing whether a company takes the plunge or holds back from adopting automated techniques.

And finally, even ostensibly international technologies like software standards or wireless protocols get modified as soon as they "touch down" at the level of the nation, the state, or the region. The global gets adapted for the local. Think, for example, about the way that core internet technologies like IP blocking and DNS filtering were reworked into censorship tools to form China's Great Firewall. These technologies, together with government policy, constructed a distinctly Chinese internet that diverges sharply from the "democratic" ideals envisioned for it by the United States. Such adaptation molds a technology, making it one's own. For states, adaptation ensures that technologies match the values they ascribe to the nation. For companies, it is savvy localization, ensuring an optimum fit between a technology and a particular market. And for individuals, it is often simply a way to make things work, to overcome hurdles and get some practical benefit. "If it is to be of any use," asserted Arnold Pacey in *The Culture of Technology*, any technology "must fit into a pattern of activity which belongs to a particular lifestyle and set of values."[46] During this acquisition process, the technology itself becomes changed: certain functions are prioritized, some are tweaked to work differently, and others are scrapped entirely. Global technology is reworked, taking on the hues of its cultural surroundings.

Going further, we might note the diverse ways that different industries within a country have adopted automated techniques. Human resources, for example, has seemed to embrace robotic process automation as a faster (if problematic) way to screen potential applicants. Health care, by contrast, is still among the least digitized sectors in certain areas.[47] In the high-stakes area of medicine, AI and automation spin up a whole range of controversial issues from data collection and privacy to the use of automated decision making. The same kinds of deep problems can be seen when introducing automation into the social safety nets meant to alleviate poverty. As Virginia Eubanks has extensively documented, high-tech tools have produced devastating outcomes for many, forcing some systems to be rolled back and replaced with hybrid approaches that blend computerized processes with face-to-face interactions.[48] Health care and welfare are distinct, each with its own histories, considerations, and communities—and understanding automation in these industries means attending to these specificities. To comprehend the implications of automation in any sector, it is vital to comprehend the deep context it sits within.

Digging deeper, we could even start to examine how these technologies differ from city to city and region to region. China's economic planning, for example, has long used a "specialized towns" model,[49] employing financial and policy incentives to privilege certain kinds of technologies in certain areas.[50] To delve into the specifics of technology use, the next chapter explores two examples of automation in a Chinese region that has now become infamous: Xinjiang.

4

Automation on the Ground

AUTOMATION IS SOMEWHERE rather than everywhere. Instead of the fuzzy conceptions of "global work" offered by automation discourse, we need to zoom into the sights and sounds of automated technologies in a specific context. Zeroing in on certain instances of automation reminds us that they are not merely a technical phenomenon but a cultural one, grounded in the history, language, knowledge, and norms of a particular social setting. This chapter takes up this task, focusing on the use of automated technologies in the Xinjiang region of northwest China.

Xinjiang's "automation" of surveillance and "automation" of agriculture are perfect examples of the dark potentials that are typically overlooked by automation's bright and overly broad visions. Yet the aim is not merely to smear or taint the automation myth but to show its fundamental limits when it comes to providing insights about the future (and the present) of work. A cultural and contextual reading is indispensable for understanding these scenarios. Here, factors like productivity, technicity, and efficiency fail to provide adequate explanations. To properly grasp what "automation" is doing here, we need to add in a cultural and contextual dimension, an analysis that folds in both operations and social relations, both technical functionality and racialized histories.

"Automated" Surveillance in Xinjiang

A figure walks into the frame. The hustle and bustle of a local market in Ürümqi surrounds her on all sides, with sellers hawking their wares and buyers bartering down the prices. But the camera focuses only on her. A red rectangle appears around her face, framing her eyes, nose, and mouth. As she wanders through the market stalls, it automatically tracks her movement, shifting its coordinates to match the on-screen image. On a server nearby, a notification is automatically generated. In a few moments, it will appear on the smartphone screens of police in the area. Via this interface, they can access her file with biometric data, recent movements, and notes from previous encounters. A minute later, they appear on-screen, quickly surrounding the figure. We need to ask you some questions, they explain, we need you to come to the station. She doesn't ask how they found her; she knows. After all, this is Xinjiang.

Located in the northwest of China, Xinjiang is officially known as the Xinjiang Uyghur Autonomous Region (XUAR). It is home to 12 million Uyghurs, a mostly Muslim population who speak their own language and see themselves as ethnically and culturally closer to Central Asian peoples such as the Kazakhs and the Tibetans. In the 1990s, this unique identity germinated into a separatist movement, with moments of violence erupting at times. For the Chinese state, these acts were not just a security risk, but a kind of existential or cultural threat. The crackdown, particularly over the past decade, has been brutal. Human rights experts estimate that up to a million Uyghurs have been detained in so-called reeducation camps, where subjects deemed to be "extremists" are reformed into model citizens.[1] Uyghur births have been suppressed through the use of forced abortion, mass sterilization, and birth control.[2] And—most relevant for the discussion here—Xinjiang has become a test bed for automated security technologies, with the state deploying one of the largest and most intensive surveillance systems in the world.

The best known of these technologies is the network of surveillance cameras. Cameras in Xinjiang are part of the nationwide security system known as Sharp Eyes. This system, proclaimed the state, would consist of half a billion cameras that were "omnipresent, fully networked, always on and fully controllable."[3] Reporters from the *Washington Post* have documented how technology giant Huawei and facial recognition company Megvii teamed up to create an "an artificial-intelligence camera system that could scan faces in a crowd and estimate each person's age, sex, and ethnicity."[4] Their collaboration resulted in a chilling new piece of functionality: "Uyghur alarms."[5] By deploying this technology, companies or authorities could receive an automated alert whenever people of this ethnicity purchased a product, entered an area, or passed through a checkpoint. Several other major camera companies have now developed similar capabilities, offering them to clients as a "feature."[6]

Technically, this is certainly achievable. In fact, computer science papers have openly celebrated their successes in "solving" this problem. With zero consideration of ethics, a group of Chinese engineers threw themselves into the challenge of creating accurate ethnic facial detection models for Uyghur, Tibetan, and Korean groups.[7] The model first extracts an image region from the face that includes the eyes, mouth, and nose. By focusing on these details, the model is able to "discover ethnic facial features," identifying key morphological traits that indicate whether an individual is within a targeted ethnic group. The model then trains on a sample set of known subjects, gradually becoming better and better at its task, until it can pick out "ethnic" individuals with high accuracy.

Yet if cameras are key, they are only the most visible component of a far more expansive regime of monitoring and control. Since 2013, every new cell phone bought in the region has to be registered and linked to a real name.[8] Since 2017, every vehicle must have a GPS tracking device installed so authorities can track their movements.[9] Wi-Fi sniffers have been set up that capture packets traveling across local networks, aiming to identify suspicious or illicit traffic. And then there is the network of checkpoints, outfitted with digital scanners, that are located around transport hubs, religious spaces, and every few hundred meters on the street. "We will implement comprehensive, round-the-clock, three-dimensional prevention and control," proclaimed the party secretary of Xinjiang; "we will resolutely achieve no blind spots, no gaps, no blank spots."[10]

Pulling all of these components together is the Integrated Joint Operations Platform (IJOP). Driven by the mantra of one platform to rule them all, this system "gathers together surveillance about the residents of Xinjiang, stores it centrally, and uses it to make automated policing decisions referred to in the database as 'pushes.'"[11] Data leaks have revealed the vast scope of this collection, a database comprising 250 million rows with a total file size of around 52 gigabytes. Where does all this data come from? The IJOP pulls from a wide variety of information sources. Wi-Fi sniffers collect the unique identifying addresses of computers and phones when they connect to the internet. Biometric data on individuals such as height, weight, fingerprints, iris scans, and blood type are included in the system. A mobile app that Uyghurs are forced to install tracks movement and other digital activity. And the formidable array of surveillance cameras operating throughout the region also feeds into this deluge of data.

Surveillance stresses that we pay attention to the purpose of automation—its content and intention. Automation discourse often assumes that any form of labor saving is an obvious benefit. But what kind of work is being automated, and how is it used within a certain context? Automation rhetoric, with its abstracted and desituated framing, its view from nowhere, rarely asks these basic questions. Its obsession with efficiency produces a blinkered focus on acceleration and optimization. It matters not what is being done, only that it is being done faster and "better." But as F. G. Jünger observes, this single-minded focus on efficiency evades the key question: To what end?[12] What is the purpose of these automated technologies? What are they accomplishing, and whom does it serve? Answering such a question requires going beyond functionalist thinking to consider the human cost exacted on specific lives and communities. Tools of optimization can also operate as tools of oppression.

From Automated Racism to Systemic Racism

These always-on technologies have been termed a form of "automated repression"[13] or "automated racism."[14] In some ways, this framing makes sense. The best way to operationalize systemic racism is by using a system. And so Xinjiang's anti-Uyghur project leverages a full spectrum of sensors, algorithms, machine learning models, automated processes, and scheduled tasks to transform hostility into a routinized activity. By embedding racist logic into combinations of hardware and software, this hostility is able to be performed incessantly, every minute, of every hour, of every day. Like a processor chip churning through instructions, it runs ceaselessly and unthinkingly until it is halted. This is how hate is made automatic.

So this technical regime certainly introduces novel conditions. Without a doubt, it shapes the lifeworlds of Uyghur people living in the Xinjiang region. For them, life today is very different than it was twenty or even ten years ago. And yet as this book has argued again and again, to suggest that this program is self-running, that machines carry out all of this on their own, would be a mistake.

For starters, much of the qualitative information fed into IJOP comes from in-home visits. "Investigative missions" are constructed by the platform and sent directly to officers, each containing instructions about the person to follow up on and what kind of information is desired. To carry out their mission, officers visit each target at home or at work, interrogating them about their religious practices, their social connections, their personal finances, and even their dietary habits. Responses to questions are filled out by the officer and then submitted to the IJOP system. These in-person interviews are long and labor intensive. "One official lamented that many colleagues have 'worked so hard' to meet the IJOP's appetite that 'their eyes are so tired and reddened.'"[15] These insights suggest that IJOP is less about automation than integration, becoming a kind of technical glue that is able to bring together a vast array of disparate sources into a cohesive platform. Here GPS data from mobile phones sit alongside in-person notes about head coverings and news from relatives. Machine-generated metadata and human observation come together to form a fuller portrait of incrimination.

On top of these visits, field research from these sites has revealed that these technical systems rely on a vast army of human labor.[16] Shops and stalls must be constantly inspected by police. Networks of neighborhood informants are drawn on to share secrets about their fellow residents. Lab technicians capture biometric data through Physicals for All, a massive program that requires every Xinjiang resident between ages 12 and 65 to provide a DNA sample. And the dense network of checkpoints throughout the region must be staffed from morning to night. "There are security checkpoints every 500 meters along all streets in Ürümqi," one police officer noted, "each guarded by over two dozen police officers."[17] All of the information created and captured during these processes must be inspected for suspicious signs and irregularities if it is to be of any use. Who does all this work? Darren Byler, who conducted years of ethnographic study in Xinjiang, estimates that all of these assessments and activity checkpoints require the deployment of more than 90,000 police officers and more than 1 million civil servants.[18]

Rather than some new "automated racism," these interventions provide a way to operationalize an already-existing racism forged from a cultural understanding and upheld through state policy. The best way to describe the relationship between modern China and the Uyghurs is colonial, argues Sean Roberts. "Modern Chinese states have clearly distinguished the Uyghurs and other local Turkic people as fundamentally different from and inferior to the dominant Han population."[19] Whether this colonization takes "softer" forms of assimilation or starker techniques of oppression, the state has never seen these ethnic minorities as equals. Systemic discrimination against the Uyghur stretches back decades, producing a common set of prejudices and a deeply ingrained distrust of this Turkic minority.[20] Seen in this light, the oppression of the Uyghur is not some unprecedented new chapter, but instead part of a historical pattern of state-sanctioned oppression of minorities that can be witnessed repeatedly. "The plight of the Uyghurs since 2017 is reminiscent of the fate of indigenous populations in the context of settler colonialism elsewhere in the world during the nineteenth and early twentieth centuries," argues Roberts, "only with twenty-first-century tools of electronic surveillance to assist in enforcement and coercion."[21]

These technologies seek to sort Uyghur. As James Leibold has documented, the state uses this multilayered network of surveillance to filter its citizenry into those who are "normal" and "trustworthy," and those who are "deviant" or "abnormal" in their thoughts or demeanor.[22] In this context, it's clear that technologies certainly play a powerful role, framing the subject in certain ways and establishing certain logics. Such algorithms, notes Louise Amoore, work to define "new thresholds of normality and abnormality, against which actions are calibrated."[23] There's no question, then, that digital technologies provide a new flexibility to this dividing line, enabling criteria to be combined and assessed in almost unlimited ways. By altering the data that is drawn on, the weighting of different variables, or the rules to be implemented, authorities can alter this boundary line to a fine degree.

But if digital technologies are undoubtedly influential, it also seems vital to acknowledge that most atavistic technology: race. Race functions as a difference-making machine, establishing a dividing line between majority Han and minority Uyghur. Race attaches traits to these human classifiers, establishing one as the modern model citizen, the other as a dangerous anomaly, an outlier bound up in religious beliefs and outmoded ways of life. The former is superior, the latter inferior. And it is this deeply embedded racialization that ultimately drives the state's use of oppressive technologies on ethnic minorities. In this sense, automated logics, while offering powerful and flexible new forms of decision making, are ultimately a way of wrapping racist beliefs in modern, data-driven tools. Digital architectures are tweaked until the boundary line between normal and abnormal suits the needs of the party. These "empirical" results confirm what the state has known all along: the Uyghur is a deviant subject who must be brought in line.

Casting the Uyghur as abnormal unlocks a broad new array of state capacities. After all, his existence means that the typical checks and balances have failed, that normal measures have proven unsuccessful. This is the incorrigible subject, who "calls up around him a number of specific interventions over and above the customary and family techniques of training and correction," noted Foucault, "that is to say, a new technology of rectification, of supercorrection."[24] Abnormality, then, gives the state license to unleash interventions that are more pervasive and penetrating, allowing it to use more drastic or draconian techniques to break this resistance and remake the subject. To deal with this extraordinary subject, the colonial power will need to take extraordinary measures.

The mass surveillance of Uyghur, then, cannot be understood without a contextual lens. This is a historical, racial, and cultural phenomenon as much as a technical one, with one people group attempting to dominate or assimilate another. To grasp this dynamic, it is necessary to grasp the state's drive toward a monocultural society. Harmony will only be achieved when the nation is transformed into a homogeneous people. In the mind of the party, diverse cultures and differing religions are a roadblock to this goal, preventing the nation from becoming a cohesive body, a "like-minded" unit. This drive toward cultural assimilation is referred to as "ethnic fusion" in government documents, but with Han Chinese representing over 91 percent of the population, it is more like overwriting or deleting rather than fusing.

This massive project is ideological or even psychological. President Xi Jinping has stressed that economic programs will not be enough to resolve ethnic tension in Xinjiang: what is needed is an "ideological cure, an effort to rewire the thinking of the region's predominantly Muslim minorities."[25] The first step is to disconnect subjects from their cultural heritage, to separate ethnic minorities from their native language, ancient customs, and traditional values. As government documents stress, reeducation should seek to "break their lineage, break their roots, break their connections, and break their origins."[26] While the reeducation camps are exhibit A in this program, its reach is both broader and more pervasive, extending into the nooks and crannies of everyday life. Breaking origins means razing houses in the name of city planning; it means framing clothing and customs as "expressions of extremification";[27] it means banning the Uyghur language in some departments and schools; and it means standardizing dress in the name of modernization.[28] All of these efforts aim to carry out a double move, destroying an established group identity and its distinctive way of life and replacing it with something new. By sanitizing and homogenizing the cultural identity of their targets, these policies and programs seek to remold them into a form more acceptable to the Chinese state. For James Diebold, "it's an attempt to re-engineer what it means to be a Uyghur."[29] Part of this reengineering project involves a program of coerced labor, and to follow this thread, the next section moves out into the fields.

"Automated" Harvesting in Xinjiang

In the background, a sea of white dots stretches to the horizon. In the foreground, a reporter begins his story. "This is the harvest season for cotton in the Xinjiang Autonomous Region," he explains. He begins moving through the rows, bending down to pluck the fluffy wisps from each plant and place them into a plastic sack. After a few minutes of work, the reporter straightens up, puffing from the exertion. "It's really a tiring job to pick this cotton by hand," he admits, "and already I felt pain in my waist and back."[30]

But luckily, he reassures the viewer, much of this process, from picking to planting and harvesting, is "now fulfilled by machines in Xinjiang."[31] The industry, we are encouraged to learn, "needs less workforce, as more and more cotton-picking machines are used instead."[32] The report cuts to a wide shot of a massive harvester driving through a field, voraciously devouring six rows of cotton at a time with its teeth, which are rapidly fed into three intake chutes on the front.

"Increasingly mechanized and smart cotton harvesting and planting has increased the sector's efficiency" notes the narrator.[33] The report cuts to a panoramic view of a cotton field with a drone buzzing overhead. The drone slowly pans across the field, emitting a fine mist of pesticide from several jets mounted underneath it. The drone, we are told, is produced by a local agricultural technology start-up. Donned in a polo shirt and jeans, the start-up's founder appears on-screen. "As growing cotton becomes more concentrated, mechanized and digitized," he explains, "Xinjiang's large cotton production is expected to transform into a high-tech cotton industry."[34]

The video cuts to a laptop with visualization software running in a browser. We see a satellite image of roads and fields from above, with each plot surrounded by a dotted white outline. One field in the center of the screen is colored in the vibrant blues and fluorescent yellows that we now associate with LIDAR (light detection and ranging) data, the same kinds of colors you might see in a self-driving car demonstration. The operator clicks this plot, and a 3D model of the field appears on screen. The operator spins and tilts the model, demonstrating how each detail has been captured, from the slightly undulating topography to a small cluster of trees on the periphery.

The report concludes by interviewing a farmer. He has been growing and harvesting cotton for decades, he says, but things have grown much easier since then. With laborers behind him moving through the rows to bend and pick cotton, he proclaims that these days he has "less stress and manual work thanks to technology."[35] The same kinds of claims are echoed in state statistics and media. The mechanization rate of cotton picking in Xinjiang is nearly 70 percent, stated one article from *China Daily*.[36] "Xinjiang cotton production has already been highly mechanized," declared one party official for the region in another article, "and even in the busy picking season, there is no need for a large number of cotton pickers."[37] New forms of automation, it seems, have been a game changer, improving efficiencies and lowering labor costs.

Human Pickers, Human Pain

Yet these claims of full automation are a fable. To interrogate them, we can turn to a report by Adrian Zenz, an anthropologist who has studied the region for years. "Despite increased mechanization," he writes, "most of the cotton produced in Xinjiang is still picked by hand."[38] The immense outpouring of labor needed to carry out cotton harvesting is coordinated by the Xinjiang Production and Construction Corp, a state-owned paramilitary organization. While some of this work has been carried out by machines in recent years, this trend does not extend to all places equally. "In 2019, mechanized harvesting in the XPCC [Xinjiang Production and Construction Corps] regions reached a share of 83 percent," Zenz notes, "however, in southern Xinjiang the mechanized harvesting share stood at only 20 percent that year."[39] There is a lumpiness to this automation, an unevenness to its deployment.

Why is the mechanization rate in southern Xinjiang so low? Look at the grade of cotton, Zenz stresses. "Importantly, 99.4 percent of the highest quality long-staple cotton was produced in southern Xinjiang."[40] There are clear disadvantages to mechanical harvesting, and these intersect with specific aspects of the region. Xinjiang is very dry, and the plants can become particularly dusty by the time the harvest season rolls around from September to November. At the same time, parts of the region can have early snowfall, which can be problematic for harvesting. All of these aspects mean that machine-picked cotton could "contain a higher percentage of impurities, such as dried leaves, dirt/dust, infertile seeds, harmful blemishes, etc."[41] Second, a single cotton plant may contain buds at different maturities: some bolls are ready; others are too young. Ingesting all of these indiscriminately through a machine lowers the quality of the product overall. And finally, cotton picked by machines requires more refining and purifying, steps that have "always been a challenge even to this day."[42] All of this downstream processing not only results in a loss of cotton material but impacts the quality of the final product.

The highest-quality cotton requires the highest-quality harvester: a human. Handpicked cotton has clear benefits. A human laborer has a deft and discriminating touch: she will not drag in bits of the plant or other debris; she can determine which bolls are ready; and her harvest has few impurities, requiring minimal postprocessing. And yet this is brutal, backbreaking labor, with no shortcuts. Workers begin at 8:00 a.m. and continue until 9:00 p.m. at night. By the end of this twelve-hour shift, even those who are fit and young are hurting. "Every part of my body aches" testified one 29-year-old migrant worker.[43] For those working to quotas or getting paid by weight, there is enormous pressure to keep up a relentless pace. Yet the faster that laborers work, the harder it is on their bodies. "Hand-picking demands both of your hands to be fast, but in turn, your hands get hurt easily too," wrote one report, "after 3 months of cotton picking, most workers are left with injured or bruised nails and redness and swelling of the fingertips."[44] These conditions are made harder by constant bug bites and Xinjiang's climate, which veers between scorching sun during the day to freezing temperatures at night.

Cotton picking is hard labor, in tough conditions, with low pay. And yet in the harvest season, hundreds of thousands of willing workers are needed. So where does all this labor come from? For Zenz, the answer is clear: from coercive labor mechanisms. "We can estimate that Aksu, Hotan, and Kashgar [provinces] alone mobilized an estimated 570,000 cotton pickers through the coercive labor transfer mechanism," he wrote in a groundbreaking report, "other ethnic minority regions that operate labor transfer schemes would easily add tens of thousands, more likely hundreds of thousands, to this figure."[45] In these schemes, the state "mobilizes" workers, signing them onto a labor contract and then busing them en masse to other regions, often far from their villages. Once committed, they are locked in, with nowhere to go. They will need to work, eat, and sleep on-site for the three months of the harvest before returning home.

While cotton picking is in some respects the epitome of manual labor, technologies of automation and optimization are leveraged to mobilize workers. Civil servants sent out to recruit in the villages of ethnic minorities are armed with a smartphone application. Through the app, cadres record highly personal details from each Uyghur household, including income levels and employment status. "Between 2014 and 2018, Xinjiang sent 350,000 cadres to Uyghur and other ethnic minority villages," Zenz notes, "the resulting data is fed into the central 'Xinjiang big data platform.'"[46]

This dynamic resonates with the "automated" surveillance of the previous section and the in-person interviews carried out by police on Uyghur households. Here too there is a real sense in which digital technologies alter the conditions for individuals. Captured data classifies and codes subjects in new ways, allowing profiles to be dynamically assembled and forms of oppression to be continually performed. And yet these data are not produced out of thin air, but by local teams working on the ground in villages and households. Automated technologies are not so much self-acting as structuring—functioning as the technical glue that scaffolds this information, that funnels it into standardized forms, and that enables it to be cross-indexed and compared against other people and places.

Coercive labor has changed the game for plantation owners. In the past, they had to hunt for workers, assembling a labor force from across China composed of a diverse array of ethnicities. The use of Uyghur communities as a massive pool of harvesters has erased this need. Plantations simply put in their requests, specifying the number of laborers needed, and the government delivers. "In the past . . . we squatted at train stations and long-distance bus stations all day, holding up signs and asking everywhere," recalled one grower; now, thanks to the optimization of this "service," they "no longer worry about finding pickers."[47] Businesses save significant money on recruitment and long-distance train tickets. A local labor pool, composed of ethnic minorities, appears when needed. This is labor on-demand, made possible by a combination of digital technology and racist "superiority."

Poverty Transformation, Thought Transformation

The app is called the "precise poverty alleviation app," a term that slots neatly into the government's claims that these labor programs are a means of breaking the poverty cycle and lifting the poor out of squalor. Cotton harvesters are certainly paid for their labor. Yet the lucrative amounts promised and the actual payment given at the end of the season are often two very different things. On top of this, the aggressive measures used to recruit workers into accepting this grueling work can sometimes be intense. Faced with these pressures, workers may feel that they have little choice in the matter. "In a system where the transition between securitization and poverty alleviation is seamless and where the threat of extralegal internment looms large," asserts Zenz, "it is impossible to define where coercion ends and where local consent may begin."[48] Indeed, for many, this "choice" seems to be no choice at all. Government documents identify the need for hundreds of thousands of workers in certain regions and speak of "transferring all those who should be transferred"; other documents refer to pickers being "mobilized and organised," transporting them to cotton fields that are hundreds of kilometers away.

In one particularly damning news program broadcast by the state news agency, officials arrive at a village in Xinjiang advertising jobs in a province thousands of kilometers away. After two days sitting in the village square with no success, they switch to a more forceful tactic: going door to door. The officials visit one family's house, attempting to recruit their daughter. Her father pleads with them to go to someone else, saying there must be others who will volunteer. The daughter herself resolutely refuses, even under the intense eye of the officials and journalists. But the officials will not take no for an answer and keep up the relentless pressure until finally, weeping, the daughter agrees. "The government continually says that people are volunteering to engage in these programs," states Xinjiang researcher Laura Murphy, "but this absolutely reveals that this is a system of coercion that people are not allowed to resist."[49] And this coercion comes with an ulterior motive. "Although the narrative is one of lifting people out of poverty," Murphy states, "there's a drive to entirely change people's lives, to separate families, disperse the population, change their language, their culture, their family structures."[50]

For the state, this program is not simply about exploiting a labor force but about transforming a mind-set. The unwillingness of Uyghur and other groups to throw themselves into cotton picking is a clear problem, a dangerous attitude linked to their old-fashioned cultural and religious values. Each harvest season presents a chance to change this outdated and unproductive worldview, an opportunity to transform the "deep-rooted, lazy thinking" of poor, rural villagers by showing them that "labour is glorious."[51] Yet as the party is well aware, this deep reengineering of hearts and minds will not take place organically. Certain forms of instruction and persuasion will be required. One regional notice on seasonal workers laid out a program of indoctrination and supervision, where work would be accompanied by "thought education and ethnic unity education" that would steer subjects into "consciously resist[ing] illegal religious activities"—a clear reference to the practices of the mostly Muslim Uyghur.[52] Another policy document stipulated that labor pools should be accompanied by officials who "eat, live, study and work with them, actively implementing thought education during cotton picking."[53]

Beyond the cotton fields of Xinjiang, researchers have noted how reeducation camps are often placed alongside factories. This proximity demonstrates the "parallel purpose of mass employment and mass internment."[54] In the reeducation camp, workers are trained to be loyal and productive subjects, learning to speak modern Mandarin, to repeat propaganda, and to perform pageants for party dignitaries. One survivor recounted how her group was forced to watch *The Hundred-Year Dream*, a film celebrating China's economic growth and power.[55] Work provides the opportunity to cement this training, to demonstrate how far they have "improved." Workers march over to the nearby factory to turn their theory into practice, completing a shift of textile labor or garment making. Their hard labor serves the market. It serves the Motherland. But most of all, they are told, it serves themselves.

Xinjiang's non-automated harvesting highlights once again the importance of grounding automation in an understanding of culture and context. The Xinjiang region produces a vast amount of high-grade cotton that requires manual picking to maintain this quality. And the Xinjiang region is home to the Uyghur, an ethnic minority that has been systematically suppressed by the state. On the one side, then, we have a capital imperative: to maximize the value of a commodity while reducing labor costs. On the other side, we have a racial imperative: to dominate and assimilate a people group that the state sees as a threat. Putting these two factors together creates a compelling logic that "incentivizes the systematic deployment of low-paid ethnic minority workers."[56]

Digital technologies are harnessed to code individuals as potential laborers, to collect the labor requirements of cotton producers, and then to coordinate the transfer of laborers to these fields. In doing so, these technologies, ranging from apps to databases and forms, establish an interface between labor demand and labor supply. They function in a similar way to vaunted "two-sided marketplaces" like Uber and Airbnb, which deliver a just-in-time connection between demand and supply. Yet by zooming in on a specific instance of automation—particular technologies, with culturally inflected values, operating on a particular community, in a particular region—we can see how this "innovative" model can also be exploitative. Cotton harvesting in Xinjiang uses automated oppression to carry out non-automated labor.

*   *   *

Xinjiang's "automated" surveillance and "automated" agriculture are overlapping but also particular: operating in distinct ways, enfolding different actors, and prompting their own sets of questions. And this is precisely the point. Rather than a grand and global "Automation," this chapter has aimed to develop a more modest but detailed portrait firmly grounded in a particular context. The application and location of automation is key.

For technologists, researchers, and policymakers, this basic but fundamental point should prompt a shift, forcing us to localize our focus and acknowledge the social, cultural, and geographical context when considering the impacts of automation. Automation is not some all-encompassing "phenomenon," but can be understood as a complex interplay of engineers and laborers, knowledge structures and domain expertise, software systems and hardware components, business logics and technical protocols—to name just a few. This intersection of elements converges in warehouses and factories, in clinics and prisons, in software studios and welfare offices. This is a dustier, noisier version of automation, where implementation is messier and perfect performances are never guaranteed.

In descending from the lofty heights of "Automation" to this automation "on the ground," it becomes obvious that workers would speak different languages, that contractors would rely on local talent, that companies would have divergent obligations to civic and national governments, and that different industries would have different priorities. Specificity is not just about adding more detail but gives us a better grasp of the technical logics and laboring subjects of automation. And this analytical purchase helps us to more fully comprehend what is at stake—what automation is doing within a certain context and why it matters. Contrasting the vivid sights and sounds of this portrait with a universal, abstracted understanding of automation highlights the latter's emptiness, its vacuity. To speak of automation occurring every-where is to speak of nothing in particular.

PART 3

The Myth of Automating Everyone

5

Automation's Racialized Fallout

AUTOMATION IS A MYTH because the catchall figure of "the human" at the center of its claims does not exist. Over the past century, automation rhetoric has consistently evoked a generic humanity as the target of its impacts. In the 1920s, the *New York Times* asked: "Will Machines Devour Man?"[1] In the 1940s, one author warned that the machine was "gaining on mankind."[2] In the 1960s, academic anthologies featured articles such as "Man, Automation, and Dignity."[3] And in the 1980s, Leontief and Duchin investigated *The Future Impact of Automation on Workers* over a sprawling 170-page report but never specified this worker in detail and scarcely acknowledged disparities of race and gender across the workforce.[4] Across these texts, it is assumed that technologically driven change is indiscriminate, affecting us all. In fact, in automation discourse, "the laborer" tends to become a fungible object, interchangeable with any other. A worker is a worker is a worker.

If opinions about automation certainly clashed, a singular humanity was always the target of its impacts. For some, automation represented the steady march of technical progress that would bring about a brighter tomorrow for all. In 1965 management theorist Herbert Simon promised that "machines will be capable, within twenty years, of doing any work a man can do."[5] For others, automation was part of a broader assault on the human staged by technology. "It is possible, if the human being falters even momentarily in accommodating himself to the technological imperative," warned critic Jacques Ellul in 1967, "he will be excluded from it completely even in his mechanical functions, much as he finds himself excluded from participation in an automated factory."[6] Yet whether seen as a blessing or a curse, the target of these transformations never seems to be in doubt: automation's impacts will apply to all "mankind."

The same framings can also be seen when we turn to fiction. In Isaac Asimov's novel *The Evitable Conflict*, the world economy has been automated, with all decision making delegated to machines. But the narrator assures readers that there is nothing to worry about because these "calculating machines" have the "good of humanity" at the core of their programming.[7] Humanity here becomes a single unit, a united population with a shared fate determined by an overarching technology. Paul Fairman's novel *I, the Machine*, is another example. The jacket asks readers to "imagine a perfect world, a world in which no one has to work, no one has to struggle, no one is deprived of anything he needs or wants. A world in which every man can indulge his most trivial whim at the push of a button."[8] Here the marvels of push-button automation will be enjoyed by all. "No one" will have to slog through a 9-to-5 job; "every man" will have his desires fulfilled and even anticipated. When it comes to imagining technology and the future of work, everyone gets lumped together.

An earlier and more naive time, some might argue. But even Frey and Osborne's far more recent study sketches out a fairly generic worker.[9] The widely cited report breaks down occupations into routine and non-routine tasks, aiming to see how susceptible certain tasks are to automation, or what they term "computerization." But this seemingly egalitarian approach smuggles in a view of man-as-motor-function through the backdoor, blurring out the race, culture, and class of the individual. This abstraction ignores the sociocultural factors that have historically steered certain groups into certain positions. Theoretically "anyone" can code and "anyone" can drive a forklift, but it is clear that not just anyone does. Stereotypes like the white brogrammer or immigrant care worker point to the skewed demographics in certain professions, demonstrating that other forces are at play.

Across the decades, then, automation's texts have consistently invoked a grand vision, a sweeping gesture that seemingly encapsulates everyone.

In deploying generic terms like "man" and "humankind," racial, social, and cultural differences are made to disappear, blurred out by these catchall categories. Here "the human" becomes a beige body, and "humanity" is turned into a shapeless mass of labor. Automation discourse adopts what Atanasoski and Vora term a technoliberalist stance, asserting that "racial difference, along with all human social difference, is transcended."[10] Indeed, claims about automation derive much of their power from this universalizing tendency. If automation is proclaimed as a gateway to a labor-free utopia, it will float all boats, lifting humanity as a whole into a new era. If automation is derided as a slippery slope to an economic nightmare, there will be no escape from the horror that ensues. Either way, it will happen to us all.

In deploying these terms, automation's human seems to be all-inclusive, all-embracing. And yet, cautions Jodi Melamed, "contemporary racial capitalism deploys liberal and multicultural terms of inclusion to value and devalue forms of humanity differentially."[11] "The human" has never been a universal category. While appeals to mankind posit a united humanity, it's clear from the past that some are granted full human status while others are relegated to the subhuman. As Jennifer Rhee reminds us, we need to take "seriously the history of the human as one of exclusion and oppression."[12] Work is marked at all times by social difference; the history of labor is a history of racial, cultural, and sexual stratification. Some workers are privileged; others are oppressed. Some earn more; others earn less. Some are celebrated; others are subjugated. Taking up Rhee's challenge, the next section briefly turns to this history, taking note of these exclusions.

Colonialism and Automation

Automation's generic human is a by-product of its apolitical, ahistorical rhetoric. In industry framings, automation is often presented as merely a set of technologies, a suite of tools that can be applied to any context. In this sense, technical processes are strictly separated from social values and the history that grounds them. An obsession with efficiency—an inherently self-referential value—contributes to this blinkered frame. Across the industry's many success stories and case studies, it is typical for automation to define its own parameters and then work to improve them. This closed logic, as Jennifer Alexander stresses, allows "the assessment of almost any action or process on the basis of the same units and qualities it had started with and nothing else."[13] When the past does get mentioned, it is a past strictly defined by technical progress. Pundits compare the sophistication of current automation technologies against the laughable state of the art in earlier cycles. Previous attempts were too early or too crude, the argument goes—but this time will be different.[14] All of these technocratic tendencies contribute to a kind of timelessness, a phenomenon lifted out of history that ignores the long and highly contested lineage of labor that it intersects with.[15]

Against this framing, we stress that automation is situated within history. The term "automation" in its current sense dates back to 1947. Ford Motors wanted to speed up its production line, and Vice President Delmar Harder suggested that what was needed was "more automation."[16] Harder, in fact, was referring only to the need to quickly move vehicle parts between assembly line stations. Yet if his reference was rather mundane, the term was quickly taken up in much broader circles and extended to refer to the mechanization of labor, particularly where it began to replace human operators.[17]

Of course, the term itself is one thing, the broader concept another. Here one could easily reach all the way back to 1811. Beginning in that year, the Luddites, an organized group of textile artisans in Britain, began months of "machine breaking," smashing weaving frames in an attempt to bring their bosses to the negotiating table. Over the next few years, a region-wide rebellion arose, only to be violently crushed by factory owners and military force. Whether championed as labor heroism or derided as anti-tech reactionism, this early moment crystallizes the clash between human and machinic labor, registering many of the qualities that continue to define debates around automation today.

From here, we might skip forward to 1867, with Marx's foundational analysis of capital. Marx was prescient in recognizing the impact of industrial machinery on workers, highlighting how these "vast automatons" integrated laborers as appendages and forced them to keep apace. Yet Marx also foresaw the limitations of this machinery and the structural contradictions it introduced into this mode of production. From there one could trace a line through to Keynes's introduction of "technological unemployment," Noble's "social history of industrial automation,"[18] James Boggs's analysis of automation from a worker's perspective, Braverman's argument about the routinized "degradation of work in the 20th century," and so on.[19] Regardless of the genealogy, the point is that automation is deeply intertwined with labor history.

Re-situating automation in history stresses its specificity. Automation is not disinterested, but rather always already embedded within a sociohistorical context. Indeed, as Adam Greenfield observes, "to consider automation with any seriousness is to be presented with a long, poignant and richly elaborated index of our deepest longings and fears."[20] Automation is inextricably bound up with a particular understanding of what work is, who the human is, and who benefits when human labor is saved. These understandings do not appear out of thin air, but like the broader concept of automation itself, are grounded in the colonial and racialized history of capitalism. As Atanasoski and Vora remind us, the "engineering projects that create the robots, program the AI, and enhance the digital infrastructure associated with a revolutionary new era are in fact predetermined by techniques of differential exploitation and dispossession within capitalism."[21] Automation does not begin from a blank slate, but from the racially inflected values of the sociocultural context it sits within.

A rich lineage of scholarly work explores this racially stratified history. Foremost among these is Cedric Robinson's formulation of racial capitalism, a concept highlighting the fact that the "development, organisation, and expansion of capitalist society pursued essentially racial directions."[22] Robinson's ambitious project is to move through hundreds of years of history and illuminate, over and over again, the often brutal racialized difference at the heart of value-creating labor regimes. Indeed, the ability to command cheap labor was fundamental to early forms of capitalism. As Jason Moore shows, the ferocious expansion of capital both financially and territorially during the long sixteenth century (1450–1640) was possible only through this constant process of "identifying, coding, and rationalizing cheap natures."[23] Anything deemed natural was a free gift, and race was instrumental in this regard. For colonists, it provided a mechanism that usefully delineated between the full human subject, with all her rights and responsibilities, and a *naturales* (natural slave) external to this category. The indigenous laborer was transformed into a natural resource, an asset that could be discovered and exploited for free. "At the heart of modernity's co-productions," notes Moore, "is the incessant reworking of the boundaries between the human and the extra-human."[24]

Work itself was key to the formation of white identity and the construction of white privilege. Roediger's seminal text on whiteness explains how a white working class emerged from the historical reconfiguration of labor and the shifting status of different races.[25] During the period of chattel slavery in the United States, the white worker could enjoy the formal distinction between his waged work and the unwaged labor of the slave. In this sense, freedom obtained much of its sweetness through its direct contrast with the bitter unfreedom of slavery. "Nothing highlighted freedom—if it did not in fact create it," stresses Toni Morrison, "like slavery."[26] When slavery was abolished, however, this dynamic shifted. Waged work was no longer the exclusive domain of the white worker but had suddenly become available to all (at least in theory, if not in practice). "White Americans could no longer derive 'satisfaction' from defining themselves as being non-slaves and non-Blacks" explains Pegah Moradi, "the greater strides made towards formal labor equality, the more Whiteness had to distinguish and elevate itself."[27] Even as employment aimed to level the playing field, whiteness pursued an invisible but tangible sense of superiority, developing and maintaining a set of systemic advantages. Everyone might have been a worker, but not everyone was treated—and paid—in the same way.

Smarter Work, Whiter Work

The latest chapter of this labor history is a push to make work "smarter." From the smartphone to the smart city, smartness has become attached to an increasing variety of devices, processes, and spaces. Smartness derives much of its desirability by contrasting itself to the dusty and clunky relics that have gone before. Those are legacy systems, old-fashioned ways of operating—or most bluntly—"dumb" products and services. Take the imaginary set out by the "smart warehouse" for example.[28] Because of its heavy workload and low efficiency, the traditional warehouse cannot match the pace necessary in today's environment. The industry's solution is to work "smarter rather than harder," using "intelligent" interconnected digital systems to undertake rote tasks within the warehouse.[29] Already here the key tenets of smartness become clear. To be smart is to embrace change, identifying faster, easier, and more efficient ways to accomplish the same tasks. To be dumb is to pass up on the opportunities handed out by the technological future, consigning yourself or your business to the slower, cruder, and more error-prone past. In the high-pressure economy painted by automation discourse, such a move lies somewhere between masochism and market suicide.

The smart vision, like automation itself, seems apolitical—a natural, commendable goal seeking only to streamline logistics and improve efficiencies. Yet smartness comes embedded with a set of values about what work should look like and who should carry it out. As critical race theorists point out, smartness has always been a highly racialized trait.[30] Sylvia Wynter has traced the historical versioning of the human, pointing out the work of deep-seated stereotypes such as the "savage" Indian or the "subrational" Negro.[31] All of these figures accomplish the same move: diminishing the cognitive abilities of non-whites while elevating the intellect of whites. More than mere bigotry, these tropes planted a flag in the cultural consciousness, claiming reason as an exclusively Caucasian domain and creating a long-standing link between whiteness and intelligence. Building atop this history are projects of scientific racism, from phrenology to more recent data-driven analyses, seeking to prove the difference in intelligence between the races. Such pseudoscience aims to rehabilitate racist tropes such as the "oversexed" native paying for physical advantages with cognitive inferiority.[32]

These stereotypes are not just a relic of a colonial past; they persist in the present. White subjects are still far more likely to be considered superior in terms of abstract thinking ability, while Blacks are considered better in athletic and rhythmic abilities, a "continuing legacy" from the days of slavery.[33] And—particularly relevant here—these stereotypes endure in more recent imagery around artificial intelligence, robotics, and automated systems. Scroll through stock images tagged with automation and AI, and you'll see page after page of white-faced androids and white-bodied robots placed in the thinking pose. Noticing this trend, scholars have worked to empirically document it. After collecting and classifying hundreds of images, their suspicions were confirmed: as soon as these technical systems become anthropomorphized through plastic molds, avatars, or 3D renderings, their facial characteristics and skin tone become white.[34] If intelligence is the domain of the white sub-ject, then making work smarter means making work whiter.

One by-product of this fallout is that minority laborers are seen as tran-sient, increasing their precarity. "Amazon's human workers are safe for now," concedes one article, "but the tide of automation is rising."[35] Implicit in this framing is that the black and brown bodies of warehouse employees are vestiges, relics of a former industrial age. They are needed for the time being but will vanish in the near future. This narrative recalls the famous folk tale of John Henry, who responded to the automation of railroad work by challenging the machine to a pile-driving competition. By summoning a herculean effort, Henry just edges out the machine, winning the battle—but then dies immediately of exhaustion. Henry becomes "the last of the steel driving men," the iconic but final figure from a line that is ultimately doomed to extinction. "The notion that there is an irreconcilable incompatibility be-tween Black labor and advanced technology—and that the latter is destined to displace the former—has been one of the most insidious and damaging of American racial myths" noted George Fredrickson forty years ago.[36] These dynamics underscore once more the uneven impact of automation. Despite its evocation of a universal humanity, automation dovetails with a colonial history in which laborers of color and mechanization were often seen as antithetical. As technologies improve and automation advances, the argu-ment goes, it is only a matter of time until these bodies disappear from the workplace altogether.

Together, these insights provide a corrective to the historical amnesia that tends to characterize automation discourse, in which the past chiefly consists of cruder technologies and the future improvements on them. Automation does not take place in a postracial vacuum, but rather builds atop a history of racial capitalism in which the stratification of labor was a feature rather than a glitch. Accumulation requires "loss, disposability, and the unequal differentia-tion of human value," asserts Melamed, "and racism enshrines the inequalities that capitalism requires."[37] While automation evokes "the human" and often claims an apolitical and "merely technical" status, this history means that its remaking of labor is always already racialized.

Automation's Racialized Fallout

Based on these uneven historical conditions, automation's fallout will also be uneven. Automation rhetoric, as the first section showed, evokes an all-encompassing humanity when it speaks of its impacts. Blanket terms like "mankind," "labor," and "worker" assert that whatever the outcome, it will happen to us all. But the mythic human at the heart of automation's claims masks the fact that some humans will benefit enormously from these transformations while others will greatly suffer. These disparities are unfolding along racialized lines. What makes this impact so unequal?

First, when it comes to automation, the ability of a worker to be resilient to technological change is directly related to that person's financial security. Here we are talking not just about income but about wealth. In *Black Wealth/White Wealth*, Oliver and Shapiro define wealth as a "command over financial resources" accumulated across generations that can be used to "create opportunities," to "secure a desired stature," and to "obtain access to life-chances."[38] In this sphere, inequality reigns, with communities diverging hugely in the resources they can draw on. In the United States, for instance, there is already a historic imbalance when it comes to race and wealth. Indeed, echoing the previous section on colonial history, we can note the correlation in this disparity. As Oliver and Shapiro stress, "the accumulation of wealth for some whites is intimately tied to the poverty of wealth for most blacks."[39] Disadvantage for some translates directly to advantage for others. Because wealth is cumulative, building up or down over time, these historical dynamics have long-term effects, producing ripples that carry forward over decades. Inequality persists in the present, maintaining racial lines. "We're sitting in a city right now," noted Anne Price in a discussion on black lives and automation, "where the average white family has around $200,000 and the average black family has $8."[40] When it comes to wealth, workers do not start from an equal playing field.

For workers, wealth offers a buffer from the shock waves of automation. Wealth allows a worker to borrow money from parents or relatives. Wealth enables a worker to survive for weeks or even months after being made redundant. Wealth can be funneled into retraining, allowing a worker to transition into a more stable and sustainable career. A lack of wealth, by contrast, exacerbates an already precarious situation, leaving individuals with little financial cushion to weather the storm. With no fallback, a worker might be railroaded into a demotion or be forced to take a bad job just to keep the lights on and the rent paid. Based on these factors, one report suggested that automation may be "widening the racial wealth gap between African American families and white families in America."[41] Sharpening our definition of "the worker," then, highlights deep, racially inflected inequality. And this inequality means automation's fallout will also be racially inflected.

If wealth is uneven, so are labor demographics. Automation is unfolding in industries that are already racially stratified. Automation has long promised to replace jobs that tick any of the "4 Ds": dull, dark, dirty, dangerous.[42] But it is not just anyone who performs this kind of labor. Historically, as sociologist Mignon Duffy notes, "dirty people" do the dirty work.[43] Indeed, despite the massive transformations in the past from servitude to service, the link between dirty work and women of color has remained consistent.[44] Humanity is not sprinkled evenly across all the available employment positions. What we see instead is "occupational crowding," argues Price—the "unnatural packing of people into certain jobs."[45] When we look at the past, we can witness a double move taking place: people of color were excluded from certain occupations and relegated to others. For workers in the automotive industry, for instance, it was clear that the adoption of high-output technical systems was accompanied by highly selective hiring practices. As the assembly line sped up and labor pressures increased, African Americans were funneled into these high-intensity roles. "In Detroit, they haven't got automation but negromation," quipped Malcolm X.[46] Accelerated and mechanized work was matched with a labor force "best suited" to take those conditions.

Occupational crowding can also be seen historically in the dirty and dangerous work of chip production. Assembling processor chips in Silicon Valley throughout the 1970s and 1980s meant running wafers through chemicals and doing chemical testing. Workers typically had minimal safety equipment, and in fact, the "clean rooms" in these facilities were designed to protect the chip, not the worker. Workers were forced to be in daily contact with an array of highly toxic compounds. These chemicals attacked the bodies of laborers over time, with workers citing headaches, dizziness, nausea, and a number later going on to develop reproductive problems. In this context, mothers, woman of color, Asian immigrants, and other marginal groups were specifically chosen by electronics management as a more pliable workforce, "socially and culturally compliant, less likely to agitate for benefits, more physically adaptable to monotonous and intricate labor tasks, and easier to control."[47]

The same patterns apply when we turn to the warehouse sector. In this notoriously dangerous industry, women, immigrants, and people of color make up significant portions of the workforce and will be disproportionately impacted. Amazon's own statistics reveal that women comprise 42 percent of employees globally, yet only 27 percent are in management; similarly, 26.5 percent of the workforce identify as Black or African American, but only 8 percent hold managerial positions.[48] Indeed, one review of the company's workforce found that 85 percent of Amazon's Black employees work in "unskilled" warehouse jobs.[49] On a broader industry level, workers of color constitute 66 percent of warehouse workers, even though they account for just 37 percent of the total U.S. labor force.[50] These numbers exemplify the "unnatural packing" of particular populations into particular roles, many dangerous or precarious. And they mean that the drastic reorganization of warehouse labor initiated by automation technologies will impact these demographics most directly. The Black sorter on the factory floor is exposed to the full physical and psychological brunt of automation, while the white manager in the office above largely witnesses these effects from afar. Both groups may experience automation, but they experience it very differently. Such deep disparities are never grasped in automation discourse and its seemingly all-encompassing category of the human. It is not "humanity" whose jobs will be replaced but particular humans.

Injuries, Activism, and Termination

What does this exposure to automation look like? One of the by-products of automation is a spike in injuries. Certainly the deployment of automated guided vehicles in some warehouses has meant that workers no longer have to walk miles per day, one of the chief complaints of early working conditions in fulfillment centers.[51] Pods of items are now delivered to workers at their station, where software prompts them to pick or stow certain objects. And yet based on this more productive process, workers are also expected to be far more productive. At one warehouse, quotas for staff have increased from 100 items per hour to 300 or even 400 items per hour.[52] The foot and hip injuries that used to plague workers have been replaced with back injuries from hurriedly lifting heavy items from pods.[53] Automation equals acceleration, drastically ramping up the pace of work and the performance targets that must be matched. The introduction of automated systems brings with it "ruthless quotas" and "dizzying speed," a combination that led to rates of reported injuries quadrupling in one warehouse from 2.9 per 100 workers in 2015 to 11.3 in 2018.[54] Far from alleviating the pressures on workers, this shift to automated systems—many ostensibly for the benefit of employees—has actually intensified it.

For these workers, automation is not some abstract discussion point, but rather a concrete shift in work practices that takes a tangible toll on their bodies. Automation is felt internally, a musculoskeletal breakdown that causes sprains, aches, and tears. And alongside these physical injuries are forms of cognitive pressure and mental anxieties created by quotas and metrics. The worry for many workers is that these "minor injuries" will gradually grow worse to the point where ankles, knees, and hips cannot carry on any longer, cutting off their income source. Rather than stop work entirely and recover fully, workers often ignore these injuries, nursing themselves back to a state good enough to make it through the next shift. In fact, this response is implicitly encouraged through various incentive schemes at the company. Keeping injuries off the books is key to maintaining insurance rates and avoiding inspections—and critics have alleged that Amazon has systematically underreported its true number of injuries, concealing the full extent of its dangerous work conditions from the public.[55] Automation pushes the pace; workers are expected to push through the pain.

In a recent letter to shareholders, CEO Jeff Bezos seemed to acknowledge this damage for the first time. "Despite what we've accomplished," he admitted, "it's clear we need a better vision of our employees' success."[56] The company had long been known as the Earth's Most Customer-Centric Company, he reminded them. Now it would also be known as the Earth's Safest Place to Work.

His solution? More automation. "We're developing new automated staffing schedules that use sophisticated algorithms to rotate employees among jobs that use different muscle-tendon groups to decrease repetitive motion and help protect employees from MSD [musculoskeletal disorder] risks," Bezos stated.[57] In essence, software would track which roles workers were assigned to and shuffle them around the warehouse to prevent them from repeatedly straining the same muscle groups. Yet this intervention merely slaps a technical bandage on the broader problem. This micro-automation fails to address the core value that drives the broader automation of work in the first place: speed. Speed is the business proposition of the company, promised through offerings like same-day delivery. Speed is thus the value embedded deep at the heart of logistical technologies in the warehouse. And it is speed, constant and unrelenting, that grinds down the muscles and tendons of workers until they cannot go on. In this sense, such technical "solutions" will always be a bandage, neglecting the gaping wound of antiworker, antihuman logic that pervades this contemporary labor regime.

One of the highest injury rates among Amazon warehouses is at the MSP1 Fulfillment Center in Shakopee, Minnesota. This area is home to one of the largest East African communities in the United States, many of whom are Somali Muslims, and much of MSP1's labor force comes from this contingent. Workers at the facility were already under heavy pressure to meet quotas, and their religious and ethnic background has intensified these pressures. Devout Muslims pray five times per day, a religious freedom that their employer is required by federal law to respect. And yet there were no prayer rooms in the facility, with workers having to use the coffee room or the work floor itself. In addition, workers say this time was clocked, meaning that for every minute they faced Mecca, their metrics got worse.[58] In the lead-up to Ramadan, workers worried that the fasting required throughout the day would make things worse. They requested that quotas be eased to allow them to participate in the holy month without exhaustion.[59] Management stated it would reduce its quotas, but the month began and these promises were never met.

Left with little recourse, workers began organizing for a strike. On December 14, 2018, at 4:00 p.m., a group of employees clocked out, bundled into hats and gloves, and then walked out the doors and into the chilly Minnesota air.

As they exited the doors, they heard a resounding cheer from the far side of the car park. A large group of off-duty employees and community advocates had been waiting for this moment and come to show their solidarity. The groups converged, holding up signs saying "Respect the East African Community" and "We're Human, Not Robots."[60] The event marked the first coordinated industrial action at an Amazon facility in North America. Since then, workers have staged several more protests and walkouts, forcing management to come to the table to discuss labor issues. Building on this inertia, workers have established a community organization, the Awood Center—"awood" being the Somali word for power. Based in Minneapolis, the center works with the 100,000 immigrants around the Twin Cities area, educating, organizing, developing leadership, and mobilizing to improve the lives of its community.

Awood will not stand idly by and let its community be steamrolled by automation. It will not watch meekly while a voracious management and advanced technologies come together to completely dictate the pace and performance of work. And here we can see the difference between the abstract subject imagined by automation discourse and the real, living-and-breathing subjects affected by these changes. The former is suggested by those generic terms like "humanity" or more labor-centric language like "workers" or the "labor pool." This is a vague but always compliant figure, a model user who embraces new technologies eagerly, adopts new work protocols with enthusiasm, and always ensures that her labor ticks the boxes required by management. Implicit in automation's dreams is what Fred Moten and Stefano Harney have termed a logistical population, one "created to do without thinking, to feel without emotion, to move without friction, to adapt without question, to translate without pause, to connect without interruption."[61] Smooth functioning assumes submissive workers.

If the automated subject is generically docile, the actual subject is a human with a backstory, a person who comes complete with a pulse, a mortgage, a kid, and a degree of resistance. This is not to suggest that every worker is an activist, only that every worker must deal with a range of life pressures that shape that person's ability—and willingness—to provide the kind of technically mediated performance that is desired. Workers can become tired, frustrated, and aggravated, especially when their livelihood is threatened.

Workers pushed to the edge may respond by dragging their feet, obstructing sensors, exploiting technical bugs, or acting in ways that are not machine readable. These practices may be supplemented with more traditional interventions such as work stoppages and strikes. "An automated subject would allow a fully automated society to run smoothly and frictionlessly," notes Mark Andrejevic, "whereas actual subjects threaten to gum up the works."[62] Sharpening our focus on the worker, then, highlights the antagonisms that individuals are able to introduce into these systems. Labor is not always pliant. Workers are not always willing to kowtow before the sign of "technical progress" that automation optimists hold up. When we develop a more detailed portrait of workers, we see that seamless operations are far from guaranteed.

In response to this and other forms of labor protest, Amazon has suspended or terminated a number of workers. At MSP1, both Bashir Mohammed and Farhiyo Warsame were fired—the latter for supposed "time-off-task" violations after she raised concerns about safety protections.[63] Wider afield, a number of workers, in both Amazon's fulfillment centers and its technology campuses, have been targeted or let go. There was John Hopkins, a Black Bay Area warehouse worker who posted union flyers and has since been suspended for ostensibly violating social distancing guidelines.[64] There was Christian Smalls, another Black warehouse worker on Staten Island who was fired and called "not smart or articulate" by Amazon's general counsel.[65] And there were Maren Costa and Emily Cunningham, user experience designers in Seattle who said they were fired for trying to "create a conversation between tech workers and warehouse workers."[66]

Amazon's ruthless response of firing activists and whistle-blowers has been called out even by other Amazon employees, with Vice President Tim Bray quitting in protest. "I'm sure it's a coincidence that every one of them is a person of color, a woman, or both. Right?" quipped Bray in his open resignation letter.[67] Building on the discussions above, we can observe how this coincidence is no coincidence at all, but rather a natural outworking of pressures placed on non-white warehouse labor and the broader racial logics embedded within neocolonial capitalism. Unemployment brings us full circle back to the question of the human. As Sylvia Wynter notes, the genre of the human is now defined by a "master code" that sweeps up class, gender, sexual orientation, superior/inferior ethnicities, and, most notably, breadwinner versus criminalized jobless poor.[68] This code includes race while going beyond it, insisting that waged work has become integral to our identity, and those who do not work are less than fully human. The jobless activists here—as well as the masses of non-white warehouse workers expected to become jobless through automation—check at least two boxes in the code, doubly excluding them from this human category. According to this cold logic, racial inferiority is "confirmed" by economic inferiority.

Equalizing Automation's Impact

Responding to activism with punishment and firing is a disappointing, if unsurprising, response. Industries might instead recognize the shock waves that automation brings, seismic shifts that impact some of its most precarious communities the hardest. Through their physical and cognitive labor, in both the warehouse and the software studio, workers have helped to create an efficient logistical machine that produces significant wealth. And yet that work is increasingly subsumed into automated systems, and that wealth goes to shareholders and managers rather than workers. Pushing back against automation's generic "humanity" helps us see the racialized dimension to this dynamic. "Amazon needs to share the wealth, because we've built a dynasty for Jeff Bezos," stressed one employee, "It's the same that they do to black people since they brought us over here."[69] Here we see history repeat itself: whiteness shapes the distribution of capital, drawing heavily on non-white labor while never fully compensating it.

Industry might do its part by taking seriously the racially inflected history of labor and positively working into this through education programs, upskilling initiatives, and diversity goals within sectors like software development and management. For Nicol Turner-Lee, "what has happened somewhere along the line when we have these conversations about the future of work, is that we have placed people of color in marginalized positions outside this economy."[70] She acknowledges that industries are becoming disrupted, that certain jobs are evaporating, and that platforms are enabling new forms of decentralized work. Based on these shifts, certain careers like data science have become vital but still suffer from a profound lack of diversity, locking women and people of color out. For Turner-Lee, then, it's clear that we "have to create a pipeline that's going to go into these jobs," with communities funneled into the production side of this digital economy rather than merely its consumption side.

For their part, workers, unions, and community organizations might think seriously about laying claim to a segment of the "means of production"[71] and the value it generates. What this would look like would vary from industry to industry and place to place. Responses in this vein might include reworking platforms into cooperatives, shifting machinery ownership to the collective, rewarding bonuses to workers based on their innovations, or dividing software royalties among authors. Common across these ideas is that automation could be taken up by workers, adapted into new forms, and its profits parceled out more equally. Embrace the technology, stresses Trebor Scholz, but adopt a different ownership model, splice in solidarity and democratic values, and ensure the proceeds benefit all.[72] These transformations too might acknowledge the history of racial inequality in labor, bringing in issues such as reparation. "For me part of the question of reparations is a question of ownership," argues Erica Smiley, "when we bailed out General Motors that should have been a call for reparations: so how much of General Motors do we now own?"[73] Other suggestions to support workers have ranged from a robot tax[74] to a more progressive income tax so that worker entitlements such as universal health care, family and medical leave, and income support for poor workers are not dependent on full-time employment.[75]

Of course, here the topic of a universal basic income, or UBI, is usually raised. While many variants exist, UBI's core premise is that every adult receives a set amount of money from the state, regardless of their work status. As certain occupations become more automated and the threat of unemployment grows, this scheme seems to offer a lifeline: a guaranteed minimum income that can cover rent, food, and other necessities. Given current conditions, UBI has increasingly shifted from fringe philosophy to mainstream solution in recent years, becoming a hot topic among policymakers, tech firms, and civil society organizations. Yet discussions of UBI, like those of automation, also tend to reference a highly generic worker. Indeed, part of the model's allure seems to be the "universal" in its title, with proponents offering a turnkey program where every citizen in every occupation is treated in the same way. Regardless of who you are or what you do, the solution to structural unemployment is the same: a lump sum that arrives in your bank account each week.

So UBI, for all its promise, can also degenerate into a narrow technical solution. Given this tendency, it's no surprise that Silicon Valley has been one of the most vocal proponents of the scheme.[76] Clearly there are some concrete dangers here. For one, such a scheme would let tech companies use their social contribution as a kind of free pass for unbridled expansion. For another, UBI could provide an excuse for the traditional social safety nets of welfare and health care to be systematically scrapped. Yet more fundamentally, UBI and its variants often fail to understand what work provides to workers in terms of identity and belonging. Work is more than a paycheck; it is about self-worth. Rather than accepting the "tech bro" solution of UBI, argues Price, we "need a whole new framing of agency, of promoting social freedom, of the idea of dignity."[77] Echoing the previous section, Price notes that for many people, you aren't a full human unless you're working full time. This deeply ingrained connection needs to be broken apart: dignity and work need to be decoupled.[78]

From UBI to robot taxes, then, there are many proposed solutions to automation and its impacts on "humanity." Yet grand claims made for top-down solutions for "workers" in the abstract should be taken with extreme caution. Labor is not monolithic, and what works for some people in some places will not work for everyone. As this chapter has argued, automation technologies will have an uneven fallout, impacting certain communities more than others. Any viable response will need to grapple with the particular mix of social, cultural, historical, and racial dynamics at play in certain industries and contexts. Who is being made vulnerable, how did they arrive at this position, and what are their needs? These are basic but fundamental questions often overlooked in big picture debates around automation and the future of work. "The way that it's being discussed amongst different elite communities is completely absent of people, which is kinda bizarre," observes Smiley, "we need to be thinking about the future of *workers*, and how working people are going to evolve."[79]

6

Automation's Gendered Inequality

IF THE MYTH OF AUTOMATING everyone obscures race, it also obscures gender. How will women fare with the introduction of automated technologies? On the face of it, this question veers dangerously close to the overly broad framings already critiqued. As Helen Hester reminds us, "there has never really been a single 'feminine' experience of work, given the myriad ways in which intersecting forms of oppression such as race, class, and able-bodiedness have affected the scope of women's opportunities, and their ability to enter or withdraw from the labour force."[1] Who exactly are the workers we're speaking of? What sectors are they in, and what sort of variance exists across it? And which countries and contexts are they inhabiting? These are all key qualifiers to add into the catchall-category of "women and automation." And yet this question can also be productive as a catalyst, a way to dig into the gendered dynamics operating in real-world industries. As we'll see, what is most remarkable is that this question took so long to ask.

So how will automation impact women? On this question, the studies are frustratingly ambivalent. On the one hand, women are already behind when it comes to labor equality. The persistence of a pay gap between genders, the unequal expectations placed on women in the workplace, and the added load of the "second shift" of meal preparation, household chores, and child care when women arrive home—all of these factors amplify the disparity between men and women in the workplace. Gender dynamics, like the racialized dynamics of the previous chapter, mean that the playing field of labor was never level to begin with. Women are starting from a more precarious position, impacting their ability to resiliently respond to technical shifts in the workplace.

Then there is the type of work that women carry out. One U.K. study found that women held 9 percent of highly automatable positions, as opposed to 4 percent for men, with "migrants, and lone parents (typically women)" particularly likely to hold jobs with "high automation potential."[2] Another U.S. study found that Hispanic women were most likely to work in "high-risk" occupations, with "high-risk" entailing an automation probability of 90 percent or higher.[3] And a third study found that "women, on average, perform more routine or codifiable tasks than men across all sectors and occupations."[4] Echoing the previous chapter, these studies point out that work is not distributed equally. Women have historically been blocked from certain types of work and crowded into others. A good proportion of this work, as the studies note, is low-paid and highly routine labor. This more predictable and controllable work presents low-hanging fruit in terms of technical challenge, placing it at the front of the line when it comes to the introduction of industrial and automated technologies.

On the face of it, this combination of existing precarity and higher replaceability suggests a bleak future. Already behind in pay and privileges, the female worker will also be the first to go. Yet at the same time, studies tend to place great faith in the adaptability of women to survive and even thrive in this environment. One study found "that women adapted to the new labor market realities by shifting disproportionately toward more high-skill occupations" and suggested that automation's "strong complementarities with tasks that require a high level of human interaction" mean that women's "comparative advantage in interpersonal and social skills" gives them a strong edge.[5]

The association between women and these softer interpersonal abilities is a stereotype, to be sure, but one that can also be found in feminist accounts. Sadie Plant's "Genderquake" from 1997 was prescient in anticipating the shock waves that automation will bring. In the essay, Plant points to a double movement in the West: a decline in heavy industry and manufacturing and a corresponding rise in the information and service sectors. This shift, she predicted, would mean major transformations in the labor market. The old skills of "muscular strength and hormonal energies" would become less necessary, losing their superior status; "speed, intelligence, and transferable, interpersonal and communications skills" would ascend in their place, becoming in demand.[6] For Plant, the implications are obvious. In the new labor environment, men would become the dinosaurs, unused to the pace of change and the suppleness required. Women would be the new apex predators, "advanced players of an economic game for which self-employment, part-time, discontinuous work, multiskilling, flexibility, and maximal adaptability" are now key.[7]

Ultimately, then, the studies are ambiguous. On the one hand, women occupy more precarious positions that are (supposedly) more prone to automation. On the other hand, women are (supposedly) more adaptable, able to flex skills that will be desirable in the forthcoming future of work. Such studies, produced by business schools and think tanks, tend to be framed in optimistic language, transferable to an action plan or policy document. However, we might also read these findings more critically or cynically. To be "prone to automation" is to be expendable and unemployed, an individual who may struggle to find the next job and the next paycheck. To be "adaptable" is to be able to survive in potentially worse conditions, to make ends meet in the face of overwhelming adversity. In this light, "adaptability" joins other neoliberal terms like "resiliency" and "flexibility" in placing the burden of employment squarely on the worker's shoulders while relinquishing the state and industry from any responsibility. The adaptable worker is the precarious worker who must invest in herself, spending her meager free time and income on retraining for the contemporary workplace. While some may find these shifts empowering, they should also be accompanied by a series of simple but critical questions: Who must adapt, what are they adapting for, and why?

Studies on women and automation point to a profound gap. While their verdicts may be ambivalent, their broader point is unequivocal: automation has long overlooked gender. For decades, automation discourse has evoked the broad category of humanity and labor without ever examining the implications for women in any detail. The fact that men and women are clustered in different occupations, that work is stratified along gender lines, that women are already behind in terms of labor equality—these basic dynamics were barely taken into consideration. As early as 1970, Jean Wells observed that studies on automation "cover all workers and give little consideration to any separate effects on women," forcing her to wrestle her results from generic statistics.[8] And this gap has largely persisted until the present. Frey and Osborne's highly cited 2013 study, for instance, constructs a highly generic worker, ignoring gendered disparities in the workplace. Across seventy-eight pages, it never mentions the word "female" and features the word "women" only twice—one in the phrase "foremen and forewomen," the other a paper on the industrial revolution in the bibliography. The end result is a void that is at once shocking and entirely unsurprising. Anne Hegewisch, Chandra Childers, and Heidi Hartmann are able to say that their 2019 report "presents the first comprehensive gender analysis of the potential impact of technological change on women."[9] Any serious consideration of automation, then, must begin by sharpening the definition of the worker and acknowledging the gender-based inequalities that continue to pervade work. The all-embracing rhetoric of automation obscures these dynamics, bracketing out the daily disparities faced by women in the workplace.

Breaking Down the Female Body

To examine these disparities, we can return briefly to the warehouse, picking up the thread begun in the previous chapter. The affordances of digital technologies provide new hinges for control and exert new pressures on workers. Always sensing and always calculating, these automated regimes take a toll on the bodies and minds of laborers. Here we zoom into the gendered dimensions of this fallout, the way injuries and penalties impact women workers.

The relentless pace of work at Amazon warehouses is well known. Several major exposés on fulfillment centers have concentrated on the frenetic speeds that are expected of workers, the dense regime of digital Taylorism that measures this performance, and the range of punishments doled out to workers who fail to meet these targets. These targets are applied unilaterally, with no adjustment for periods of illness, or grieving, or family crises. A union in the U.K. documented how a woman with breast cancer was allegedly put on a "performance improvement plan," while another was sent on a business trip the day after having a miscarriage.[10] After clocking in, every second counts, leading to anecdotes of workers skipping bathroom breaks or peeing in jars in order to maintain their quotas. This breakneck tempo must be maintained hour after hour, day after day, week after week.

Less highlighted is the unending repetition of this labor. One of the singular qualities of automation in this instance is its ability to combine frenetic activity with mind-numbing boredom. "My body is grudgingly adjusting to the job, but my brain isn't," reflected Emily Guendelsberger in her account of work at an Amazon fulfillment center, "I've started dreading the monotony even more than the pain."[11] In her case, the job consisted of checking her handheld scanner, walking to that location in the warehouse, picking up an item, then checking her scanner for the next product. Scan and pick, scan and pick, ad infinitum—for eleven hours straight. Ways to relieve this monotony were severely limited. Headphones for listening to podcasts or music were verboten. Conversation with other workers was banned or at least significantly discouraged. And even if it wasn't, the optimized picking routes workers took through the cavernous warehouse space meant that contact with other associates was often minimal. Work on the top floor was dreaded by workers not only because of the longer distances to walk, but because it was so lonely.

The result is a strange blend of breakneck pace and silent isolation. The work demands an absolutely grueling physical performance while also stipulating that the worker's mental and social capacities be suppressed. "Low-wage workers," observes Guendelsberger, "have a strong-bordering-on-mandatory incentive to crush those unuseful human parts of themselves down to atomic size."[12] Creativity, ingenuity, problem solving—in this position, none of these aspects are wanted. The body pounds out mile after mile, walking, squatting, lifting, and pushing through the pain, while the mind is given nothing at all to occupy itself.

For workers, this synergy of forces produces an immense and unrelenting pressure, a chain of tasks that stretches into the distance with nothing to provide relief or even variation. Alongside their physical health, their mental health starts to break down. "I cry on my way to work," admitted one employee in a post on a forum, "I hate not speaking to a single person for my entire 12-hour shift."[13] The monotony led her to stare incessantly at the clock, wishing the shift was over so she could at least go home to sleep before starting again. "My mental health is starting to affect me physically as well, on my way to work yesterday I had to pull over on the freeway to vomit." Another worker confessed that she had her "first amazonian mental breakdown at work last night. I think I've officially lost my absolute mind working here. It started with me talking to myself but everyone does that here . . . last night I just sat down on my uboat [utility cart] and cried."[14] Responses to these posts commiserate and add their own stories of depression and anxiety, slowly generating a potent portrait of worker suffering in the twenty-first century.

In some cases, mental health declines to the point where suicide is considered. One woman at MSP1 admitted she wanted to kill herself and mentioned using box cutters; another woman at a fulfillment center said she was going to stab herself in the stomach.[15] These cases slot into a long litany of mental breakdowns documented at warehouses over the past five years. At a Florida warehouse, one woman said "she was going to go home and kill herself"; a supervisor "witnessed her hurting herself because she had been fired and felt she 'did not have anything to live for.'"[16] It's worth noting here that firing at Amazon fulfillment centers is an entirely automated process. One way to be fired is to lose all of the six points you start out with. Leave your shift an hour before you're due to be finished, and you lose a point. Don't come to work and don't call in? Lose three points. But a more pervasive way is through failing to meet quotas. Whether picking or sorting, employees are exhaustively tracked and expected to consistently hit ambitious targets. In just one facility over the course of one year, roughly 300 full-time associates were terminated for inefficiency. Amazon's system monitors the worker in real time and "automatically generates any warnings or terminations regarding quality or productivity without input from supervisors."[17]

This termination system demonstrates how automation is about control as much as efficiency. "The impulse to automate," observed David Noble, was encouraged by the "continuing struggle of management to gain greater control over production, weaken the power of unions, and otherwise adjust a recalcitrant work force to the realities of lower pay, tighter discipline, and frequent layoffs."[18] To be sure, firing is faster with software that fills forms and updates rosters than with each case being manually reviewed by someone in human resources. But more fundamentally, this system lets management dictate the rules and logics of termination while removing any agency from workers. For workers who are penalized or even terminated, there is no one to negotiate with, nowhere to plead their case. When the algorithm is your arbiter, the system's decision is final: it cannot be reviewed or revoked. Automated firing, then, is a prime example of a broader imperative to restrict the political agency of workers. Technical systems seek to shrink the field of action available to organized labor, aiming to diminish the space of contestation to zero.

Suicidal ideations can also take the form of a cry for help. Amazon workers sometimes write messages on the sides of boxes around the warehouse. While these are mostly banal or even positive, one forum post documented a much darker version. Scrawled in black marker on a cardboard carton was the phrase: "Im gonna jump off the third floor."[19] One respondent slapped a white sticker below the phrase, writing "call for help" and listing the phone number of a mental health line. But other comments were not so kind. One arrow pointed to the phrase and quipped, "do a barrel roll." Another joined in on the dogpiling, scribbling "do a flip, fucker" with a Sharpie. The worker who originally posted the message is asking for help, for some kind of support from those around her. But these responses are snide and disparaging rather than uplifting.

Here we can see how the technical shapes the social, with automation creating an environment that actively undermines solidarity. Digital tracking means that every second counts, forcing the worker to keep moving. Rather than discussed over an adequate break period, the cry for help is hastily scrawled on a cardboard box. Algorithmic route planning fans workers out, ensuring that contact with others is brief or non-existent. The cry for help is not a face-to-face conversation, but a message left for others to hopefully find in the vast maze of the warehouse. And finally and most fundamentally, this automated regime is highly atomized, with stats, rewards, and punishments attached to each individual. In the digital logic employed by management, there is no "we," no sense of togetherness. In fact, gamified initiatives like

Power Hour, where pickers compete for a gift certificate and other prizes, actively set workers against each other. Rather than a community, warehouse workers are only a temporary aggregation of three thousand individual profiles. It's unsurprising, then, that the message was met more with hostility than sympathy. This mix of automated technology pries workers apart, considering each as a stand-alone unit in competition with others. Unity and communality are incompatible with this work regime. If forms of solidarity do persist, they do so in the face of this dominant logic.

These injuries echo the unequal fallout seen in the previous chapter, the way that automation's impacts fall unevenly along racial and gender lines. This dynamic underscores the need to go beyond the generic definition of "labor" offered by automation discourse. Automation does not apply to a monolithic humanity but to particular people with particular genders in particular places. Policies and plans that fail to account for gender will miss these key dynamics and the way they shape the interaction between automated technologies, human labor, and workplace practices. For organizations advocating for better work conditions, this nuancing of "the worker" also nuances the rights and demands that are advocated for. Gender and race push organizers to think beyond the paycheck, foregrounding an array of key concerns ranging from equality to discrimination and sexual harassment.

Work and Non-Work

Yet while labor conditions are certainly central and significant, focusing on women confronts us with a more fundamental question about what work is. Within automation discourse, "jobs" and the "future of work" go hand in hand.[20] Countless debates focus on jobs. Which jobs will be created, and which jobs will be deleted in the coming technological apocalypse? The same assumptions can be seen when discussing the target of automation, which is always the workplace as traditionally understood. If automation is "coming soon to assembly lines, warehouses, and hallways near you," as Brynjolfsson and McAfee warn, then the question is which workers will be left on the assembly line or in the warehouse.[21] Implicit in these debates is that work is something carried out by staff for an employer to obtain an economic benefit. Work is waged work, performed for a price. Automation discourse accepts this capitalist concept of work, a narrow definition that "has come to dominate the concept of work itself."[22]

Why should work be defined in this way? Already it should be clear that a gendered dimension is at work here, a high-stakes framing that puts women on the losing side. Accepting this conventional view of "work" privileges the labor of some while excluding the contributions of others. Pushing against the myth of automation means pushing against this narrow conception of work. In these final sections, then, I draw on feminist scholars to challenge this status quo account and its self-serving compartmentalization of work and non-work. We'll shift from the warehouse to the home, examining domestic automation and tracing how gender has deeply shaped the development of technology from early on.

What counts as work? For feminist and theorist Silvia Federici, this question is not merely semantic but highly strategic in defining the site of political struggle. Work is typically regarded as work only when certain people ("employees") carry it out in certain places ("the workplace") and receive monetary compensation ("income") for it. If the accumulation of capital and the exploitation of labor occurs only with waged workers on the factory floor, then resistance against this exploitation is constrained to a narrow arena. This is the laudable but limited work of advocating for shorter hours, higher pay, better conditions, and so on. But what happens when we reconsider this self-serving definition, when we open back up the concept of work itself?

If work is something more basic and yet larger—the broader constellation of practices Susan Ferguson refers to as "life-making"[23]—then the arena of contest also extends to everywhere those activities occur. From caring for children to tending the sick and maintaining the house, this is the vast terrain of reproductive work, the crucial but often invisible labor needed to sustain current and future workers. Clearly this is skilled work, often drawing on a sophisticated mix of cognitive, social, and tacit capabilities adapted for a specific person or context. And yet these tasks are viewed as duties or necessities, "an extension of naturally occurring feminine (and often, quite specifically, maternal) predilections, affects, modes of intimacy, personal preferences," and so on.[24] Historically these skills were developed in the home and performed by women without pay. Whether framed as supporting the family or sustaining the so-called breadwinner, this "women's work" was never elevated to the level of work proper. It is here that Marxist critiques falter. Like automation discourse, they accepted the narrow definition of what work is, limiting their scope and power. "Marx failed to recognize the importance of reproductive work because he accepted the capitalist criteria for what constitutes work," Federici asserts; "he believed that waged industrial work was the stage on which the battle for humanity's emancipation would be played."[25]

For Federici, it is reproduction, rather than production, that presents the key site of labor struggle in the twenty-first century and the one that holds the most promise for advancing the political agency of women. Rather than focusing exclusively on waged labor outside the home, reproduction shifts our attention to unwaged labor inside the home. Our gaze turns from industry to domesticity, from generating products to generating the key figure behind all those products: the worker herself. For Federici, this paradigm shift reconfigures "our image of society as an immense circuit of domestic plantations and assembly lines where the production of workers is articulated on a daily and generational basis."[26] Worksites proliferate and workers swell. Suddenly we see care work and house work, affective labor and immaterial labor. Work is here at the kindergarten or play center; it is there at the church, the mosque, or the synagogue. We see individuals working with grandparents and grandchildren, with communities and charities, with communal gardens and environmental projects. Immediately it is clear that work is taking place everywhere, all the time. Dismissing this labor as non-work dismisses its crucial contribution and marginalizes those who undertake it.

Automation's Mixed History

Automation's tight coupling with colonial and capitalist histories means that it adopts these same assumptions. Here, "work" is paid work at the office, clinic, or factory; "non-work" is the unpaid reproductive labor carried out by women that takes place in the home or elsewhere. Given these patriarchal assumptions, the history of automation becomes clear. Automation has concentrated heavily on "work," pouring vast amounts of time, research, and capital into optimizing labor at the (male-dominated) worksite, while largely ignoring what is "not work"—housework, care work, sex work, and other forms of gendered or domesticated labor.

Indeed, it is striking how untouched this sphere is by the "disruptions" of the technology sector. Automated work in the office, the warehouse, or the factory is the focus of starry-eyed futurists and gushing tech editorials. Every quarter, industry rolls out the latest "game-changing" innovations, technologies that will reinvent outdated processes and optimize inefficient workflows. While these promises often fall flat, they nevertheless demonstrate the focal point of the automation imaginary. When set against this frenzy of activity, the "non-work" of reproductive labor is still non-automated. Activities like caring, teaching, feeding, cleaning, and sustaining continue to be carried out in person, without technical augmentation or mediation. This is manual labor, done by hand, with no shortcuts. "Amidst all this prophesied social change to come, these so-solid promises, underpinned by gadget-scale instantiation, algorithmic advancement, prototypical experimentation, hard investment," observe Caroline Bassett, Sarah Kember, and Kate O'Riordan, this is one sphere of life "that often appears obdurately unchanged."[27]

In the history of technology, the home is an afterthought. The devices that did make their way into the household—the microwave, the washing machine, the vacuum, the refrigerator—did so almost accidentally. They were originally conceived for military or industrial use, those lucrative sectors firmly associated with traditional definitions of "work." Only later, sometimes decades later, were they repurposed and marketed for the home. "Given that much domestic technology has its origins in very different spheres, rather than being specifically designed to save time in the household," writes Judy Wacjman, "it is not surprising that its impact on domestic labour has been mixed."[28]

"Mixed" suggests that technology has not lived up to the hype, that its domestic benefit has been far less than advertised. Rather than removing work, automation reorganizes it—and sometimes adds more. In her book *Pressed for Time,* Wacjman demonstrated that household appliances such as microwaves and deep freezers made no impact on the time women spent doing food preparation and cleanup—and some appliances like a clothes dryer actually increased time devoted to these chores. Wacjman's findings line up with Ruth Schwartz Cowan's much earlier work on household technology from the mid-1980s. In her celebrated study, Cowan showed that new devices made some chores less physically demanding but also introduced a whole range of new tasks. This was coupled with rising standards in domestic labor and higher expectations placed on the housewife. Now she was expected to provide a rotating menu of elaborate and varied meals, to ensure that rugs and floors were spotless, to iron clothing for family members once a week—in short, to maintain their higher standards through her higher labor output. Far from erasing work, Cowan argued, technology in the home led to "more work for mother."[29]

Why was this history so mixed, failing to produce a meaningful lessening of labor? Why was the factory a focal point while the household received technology's hand-me-downs decades later? While there are undoubtedly countless factors at play here, many seem to come back to that core distinction between work and non-work. Work takes place in the workplace, a productive process that creates value for companies and for the nation as a whole. That value can be measured in terms of manufacturing quotas, or annual sales, or gross domestic product. Non-work takes place in that nebulous sphere outside the workplace. It is "non-productive" at least in part because it cannot be measured in conventional terms.[30] Who knows the myriad tasks that take place in the hidden abode of the home? And even if you did, how could you quantify a year of raising a child, or providing a consistently clean home environment, or offering a loving and supportive conversation? Without being sold for a dollar value on the market, these contributions become immeasurable. And this messy, nebulous quality means that this realm of activities is bracketed off from the economy and economic analysis.

Based on this patriarchal distinction, the development of technology becomes highly logical. Resources in the form of capital investment, research programs, government incentives, and specialized training are poured into the productive and profitable sphere of work. These resources are finite, limited, part of a zero-sum game. And so this vast concentration of money, talent, and time is simultaneously a diversion from the unproductive and unpaid sphere of non-work. The military-industrial arena, as we saw, is given ample time and capital to experiment and innovate, while the home remains an underfunded afterthought. Caring, cleaning, and nurturing continue much in the same vein as in previous generations, and this is not just a function of their non-routine nature but because of how work itself is defined. Jason Smith notes, "these sorts of activities, however necessary they may be for the reproduction of capitalist class relations, are always the last to be rationalized, that is, made more efficient, and less onerous, by means of labor-saving innovations."[31] Since at least the early 1980s, feminist scholars have hoped that automated technologies would dispense with the need for housework altogether.[32] But this mixed history suggests fundamental reasons why those dreams of a labor-free liberation will never materialize. The paths pursued by automation adopted a distinctly gendered dynamic.

A Man's Home, a Man's Castle

When we do see automation applied to the home, this technology is typically developed by and for men. Advertising for home automation has been aimed squarely at a masculine audience. Search for "home automation" online, and you'll be scrolling through countless variations of the same image: a man, dressed in a suit, with his finger poised over a smartphone screen brimming with control panel icons. Whether completing their bachelor pad or taken on as a weekend project, these are systems installed and maintained by men. In fact, when it comes to purchasing devices and choosing features, women's interjections can be regarded as interference. "Am I the only one with a wife who is suckin' the fun out of their home automation plans?" complained one man on a home automation forum.[33] You need to work on your WAF, your "wife acceptance factor," suggested the top-rated reply. Keep it simple stupid, recommended another user; your automation strategy should obey the principle of the "smart home for dumb people."

Home automation offers mastery over the domestic domain. By wiring up connected thermostats, smart lighting, media systems, and other household gadgets, the user can attain unprecedented levels of control over their living space. Through clever combinations of software and hardware, the home itself becomes a system that can be managed and manipulated. Home automation, then, comes with a distinct view of what the home is and what it needs to accomplish. To set the room temperature three degrees cooler, secure the perimeter, dim the lights, and cue the sports channel—these are the kinds of tasks that home automation strives to manage and manipulate.

Because of this focus on technically oriented men, housework is rarely included in the imaginary of the automated home. With the exception of robo-vacuums, devices to cook meals, scrub bathrooms, make beds, and carry out the litany of other everyday household chores are largely absent from the glossy brochures and convention displays of home automation companies. As Anne Berg observed: "housework has no place in the general idea of what a smart house is."[34] Of course, this is not to suggest that these tasks are exclusively women's work, but to simply observe that home automation as a vision panders to men while failing to account for women consumers, something feminist scholars have noted. "The 'smart house' is a technology rendered masculine by the process of design and marketing," notes Deborah Chambers, "a technical construction that preserves gendered power relations."[35]

One recent incarnation of the automated home comes in the form of digital assistants. Products like Google Home and Amazon Alexa are small electronic devices with microphones connected to the cloud. Upon hearing their "wake-word," these assistants listen to the user's commands and then respond by playing music, delivering news, telling jokes, ordering products, or undertaking any number of other tasks. These systems are highly complex combinations of hardware, software, and network infrastructure. Yet in slogans and commercials, this technicality is downplayed. Instead this assemblage of elements is framed as a personality, a warm and empathetic presence who is "always listening." After combing through logs, one Amazon developer was surprised to find that "every day, hundreds of thousands of people say 'good morning' to Alexa."[36] Another user admitted that "I have a very familiar relationship with my Echo. I talk to it like it's an actual person."[37] Such treatment is by design. As one study observed, Alexa is endowed with a name and gender, she can interact playfully, she is co-located with users, and she can alter the environment—all of which "are designed to afford social functionalities and promote anthropomorphism."[38] Rather than an algorithmic bundle of technologies, Alexa is experienced as an affective persona.

The goal is to make this persona warm and welcoming, affirmative and open. Alexa's product chief has said the team's aim is to develop something that is friendly, can turn off your lights, chat about anything, and empathize when you're having a bad day.[39] Alexa has now added an ability to distinguish between voices. Rather than "zeroed out" for a generic user after each use, this capability allows for ongoing relationships. Moreover, recent patents attempt to delve further into the life of the individual, detecting vocal variations that betray one's present health condition and current mood. So while Amazon is constantly updating her capacities, the overall vector is clear: to develop something chattier and chummier, something more affective and emotionally attuned. In doing so, Alexa embodies what theorist Byung-Chul Han has called "friendly power." Such power operates not through restrictive obedience but via relentless affirmation. As Han asserts, "It says 'yes' more often than 'no'; it operates seductively, not repressively. It seeks to call forth positive emotions and exploit them."[40] Always attentive and always supportive, Alexa remains structurally open to every utterance.

Gender is a key component of this identity. By emulating gender through voice and name, designers are drawing on a long history of feminized technology. One starting point might be *The Future Eve*, a nineteenth-century novel that popularized the term "android." In this misogynistic tale, a male inventor creates a machine that emulates and even improves upon his human love interest. This lineage would include "Eliza," the early psychotherapist computer program developed by Joseph Weizenbaum in the mid-twentieth century. Skipping forward, there is "Siri," Apple's female-gendered assistant for televisions, watches, and desktops. In the same vein, Microsoft's "Cortana" assistant for Windows is based on a voluptuous female character from its Halo video game series. And more recently, there is "Xiaoice," Microsoft's so-called girlfriend bot that has millions of users. Even Google Now, an ostensibly genderless voice assistant, began life code-named "Project Majel." Majel Barrett acted as a nurse on the original Star Trek series, a role revolving primarily around her unrequited love for officer Spock. Barrett subsequently became the onboard voice of Federation starships, tirelessly serving each of the crews in each of the Star Trek television series and in most of the Star Trek movies.

These assistants are always listening, always ready to serve, always willing to acquiesce to the user's demands. For Hannah Gold, these products thus share a common characteristic: "what has traditionally been perceived as female instinct, experience, and voice is artificialized, replicated, and sold."[41] They tap into a seam of servitude, a lineage of technologies that build directly on the conventions established by gendered labor. By leveraging this history, these products establish a particular relationship: the master commands, the assistant responds.

Programmed Inequality

If automated assistants draw on long-standing feminine tropes, actual women have long been sidelined from automated technologies. To dig into one particular history of gender and automation, we can turn to *Programmed Inequality*, Mar Hicks's study on women and the British computing industry from the 1940s to the 1970s. In the first decade of this period, computer operating and programming was considered women's work. Women were pioneers in this nascent space, rapidly developing expertise in maintaining these systems and performing work through them. During the war, women were integral to the efforts of computing and code breaking taking place at Bletchley Park, yet this role was largely covered up. In the years after, computing was framed as a low-skill clerical domain, ensuring women remained low paid. These women were placed on the bottom rung by management, giving them little chance for advancement and little power for negotiating.

Across the Atlantic, the same kind of dismissal had occurred with a computer for U.S. military use. The ENIAC—or Electronic Numerical Integrator and Computer—was designed to calculate bomb trajectories, calculations that could take a human twenty hours or more to work through. A machine was developed and built, but its inventors quickly realized that it would need to be reprogrammed with the use of punch cards. Six women were promoted from human computers to machine operators based on their mathematical skill. They had to familiarize themselves intimately with the machine, a 30 ton behemoth that sprawled across 130 meters and housed thousands of vacuum tubes, relays, and mechanical switches. In February 1946, journalists descended on ENIAC's home in Pennsylvania to see it unveiled. A male senior engineer showcased the machine's features, wowing the press by adding together 5,000 numbers in a second and then calculating a bomb trajectory. Implicit in the demonstration was that ENIAC was "an autonomous, intelligent machine"—a framing that reporters echoed in stories describing it as an "man-made robot brain" or a "wizard."[42] But unseen and unacknowledged in the spectacle was the gritty and groundbreaking labor of the women who had programmed it. The six women operators "who had crawled through the machine's wires and vacuum tubes to enable the so-called acts of machine intelligence" were never mentioned.[43]

As computing advanced throughout the 1950s, hopes of automation grew. While these systems still required human operators, industry dreamed these bodies would soon be done away with altogether. "Human labor was positioned as a temporary inconvenience within systems that anticipated a more fully automated future. If labor could not be made fully invisible, at least it could be made less obtrusive by using temporary, high-turnover women workers with no claim to equal wages, job training, or promotion opportunities." And yet, echoing the insights from Chapter 1, this grand vision would never be attained. In a section titled "the fiction of full automation," Hicks relates the broad set of technical and institutional stumbling blocks that hindered the smooth rollout of these systems. While industry advertisements displayed these computers in modern facilities with gleaming surfaces, the actual sites of installation were often cramped and outdated. Creaky infrastructure and aging buildings meant that basic requirements for computing—stable power and cool temperatures—were hard to come by. These sites were "almost inimical to computerization on a physical level. Vacuum tubes often broke in environments without temperature control. Punched card entry could fail due to moisture-warped cards."[44] Operators had to coax or even kick machines to keep them running. The effortless future of total automation clashed with the harsh realities on the ground.

In the 1960s, computing began to gain steam. Women had always been present as operators, but their labor was an open secret, unrecognized and invalidated. Now, in magazine ads and brochures, there was a noticeable shift, with women positioned front and center. As Hicks notes, women's labor "was being marketed explicitly as an essential part of the computing system."[45] And yet what kind of woman was this, and what role did she play? She was the "girl operator at your terminal," the pretty young thing who translated a man's commands into machine commands—the interface to an interface. Far from being empowering, her presence proved how intuitive computers had become. Here was a system so simple it could be run by a woman with basic clerical skills. For the company that invested in this automated technology, cost savings would come not only from information processing but from the cheap feminine labor it unlocked. "Women's labor was no longer simply the best fit for the system," Hicks stresses, but the "necessary element in order to get the benefits of cost and control computers promised."[46] Women were part of the package, the low-paid wetware that accompanied the hardware.

Here we see how automated technology comes together with gender and class formations to stitch up fraying norms in the modern workplace. The computer introduced new forms of labor and new roles, which were theoretically open-ended in their remuneration, conditions, and prestige—and yet sociocultural dynamics almost immediately begin to bifurcate this work, funneling women into low-paid transient positions while elevating their masculine peers into jobs associated with authority and expertise. Women's work was crucial to the complex information-processing projects that began emerging during this period. And yet they were classed as merely typists or punchers, deskilled workers who could easily be replaced with another if the conditions or pay weren't to their liking. Time and time again, Hicks demonstrates how "work identities were formed by job classifications and management language that was carefully deployed to protect a status quo of structural inequality."[47]

By the 1970s and certainly by the 1980s, the central role of information technology became increasingly clear. As Hicks argues, "greater automation would in fact require more workers—and more skilled workers—than the government expected."[48] Individuals were needed who understood the ins and outs of these technical systems and the often obscure logic of data. Operators were in the perfect position to guide the transition from traditional pen-and-paper conventions to newer practices driven by information processing. And yet from early on, industry and the state had come together to create a "deskilled feminized class of machine workers."[49] Women's labor had been systematically downgraded and dismissed. Three decades of structural inequality had successfully shifted these early insiders to marginal positions on the fringe. Women had been bracketed out of the industry in any meaningful capacity. Computing was marked off as a solely masculine domain, a field of hard logic and pure reason that women had no place within.

Today, the importance of this industry has only grown. From enterprise software to social media and gig economy apps, these are the products, platforms, and services that millions interact with on a daily basis. Because of this, the designers and developers in this space exert outsize influence, shaping the functionality that is offered, the content that is privileged, and the kinds of experiences that users have. Skilled workers in these roles are highly sought after and highly paid. Developers can quickly earn six figures straight out of college, and those with expertise in machine learning and AI systems far more than that. These jobs score high in terms of agency, earnings, and stability. They have driven a turn to sciences, technology, engineering, and mathematics—the so-called STEM disciplines. In recognition of this, schools, governments, and non-profits are all endeavoring to funnel more women into these roles. Coding camps, company recruitment drives, and policy directives are just some of the incentives attempting to reverse the male domination of the technology sector. And yet this bias is now deeply ingrained, turning these efforts into an uphill task. As Hicks notes, "initiatives to get girls, women, and people of color to train for STEM jobs cannot undo the underlying structures of power that have been designed into technological systems over the course of decades."[50]

Automated (Re)Production

While notable exceptions exist, women often remain marginalized from technical domains, with countless stories of female workers who recount being sexually harassed, bullied, or isolated, driving them quickly out of the sector. Their ability to shape technologies of automation being implemented on the ground suffers as a result. Designing and developing more egalitarian systems will need to start from this basic gender gap, recognizing who dominates this conversation and who is absent from it. In the previous chapter, we saw how automation discourse deploys terms like "mankind" and "humanity" in making its claims. These terms presume a future that has moved "past" race and gender, a world where difference has been flattened and erased. But the history of women in computing—from many histories that could have been chosen—demonstrates the stratifications and exclusion that continue to define work. Any form of feminist automation would start by acknowledging the deeply gendered division of the tech industry and actively working against it.

At the same time, we can also rethink the definition of work itself, questioning the dividing lines between work and non-work. Within automation discourse, work was waged work in the workplace. Automation often oscillated between ignoring the domestic sphere or applying masculine-designed "solutions" to it, with mixed results. Yet in our precarious and postpandemic world, the clean division between the workplace and the home has been blurred. Work takes place in the bedroom and the living room, during meal prep or the school run—in short, everywhere and anywhere. Requests from colleagues and children come through on the same platform. We juggle caring and coding, main hustles with side hustles. As Bassett, Kember, and O'Riordan suggest, rethinking automation would begin by recognizing that the habitual divisions of "internal and external, public and private, work and leisure" that define its analysis are "ideological and convenient fictions."[51]

This insight means that work cannot be bracketed off. If work is pervasive, our critical engagement with automation must be equally so. Automation shapes production and reproduction, industry and domesticity. To properly address these tangled conditions, a broader and more overarching paradigm is needed. One example of such a holistic approach is what veteran labor activist Jane McAlevey calls "whole-worker" organizing. For McAlevey, "whole-worker organizing begins with the recognition that real people do not live two separate lives, one beginning when they arrive at work and punch the clock and another when they punch out at the end of their shift."[52] Work spills over into (supposed) non-work, and what happens in the home and school feeds back into the factory or office. This overlap can be taken up and employed strategically by those seeking to improve the lives of themselves and their community. Indeed for women, it is automation's work at this boundary point, this messy intersection, which may be of most interest.

How should we respond to the automation of life at this conjuncture? There can be no pat answers, no hard-and-fast rules. Each context presents a unique set of issues that need to be considered. Each technology introduces a double-edged mix of potentials and pathologies that have to be negotiated. We need a fine-grain distinction, stress Helen Hester and Nick Srnicek, one "attentive to the nuances of specific technologies; to questions of access, ownership, and design; and to the way in which ideas of gender and work become embedded within the affects we associate with technology."[53] Rather than accepting or rejecting "automation" in all caps, this means zooming into each community and context, exploring different versions, examining their sociomaterial impacts, and weighing up their relative worth. This is how we start to undo the myth of automation.

CONCLUSION

Automation Is Not Our Future

AUTOMATION, WE ARE TOLD, IS OUR FUTURE. Putting together the triple fictions of the previous chapters—full autonomy as a global phenomenon that will impact everyone—leads to the conclusion embraced by the automation optimists: automation is the dawn of a new era. "We are convinced that we are at an inflection point," intone Brynjolfsson and McAfee bombastically, "the early stages of a shift as profound as that brought on by the Industrial Revolution."[1] Some call this the Information Age; others refer to it as the Second Machine Age or the Automation Age. Whatever the term, this is a new epoch, a period of dramatic change that slots into a grand procession running from the Bronze Age and the Iron Age to the Industrial Age.

According to this pseudo-historical account, each of these periods was defined by major technical advances. Traditional ways of life were disrupted by technological progress. Suddenly existing tools and weapons became inferior; conventional modes of production were rendered inadequate. In this tale, technological adaptation presented a litmus test: those who survived and thrived embraced these new technologies, while those who rejected them were a dying breed. Technological change was inevitable; it was only a question of who recognized this and who refused to. The moral in this story is clear: the winners got with the program; the losers bucked against it.

For its proponents, the Automation Age ushers in the same imperative. Progress is marching forward, and these changes are coming whether you like it or not. In his slim masterpiece 24/7, Jonathan Crary critiques the framing of technology as an age. "One of the consequences of representing global contemporaneity in the form of a new technological epoch is the sense of historical inevitability attributed to changes in large-scale economic developments and in the micro-phenomena of everyday life," he writes, "the idea of technological change as quasi-autonomous, driven by some process of autopoesis or self-organization, allows many aspects of contemporary social reality to be accepted as necessary, unalterable circumstances, akin to facts of nature."[2]

There is a kind of fatalism to this framing where automation—together with the relations and practices it imposes—becomes inescapable. As Crary suggests, the metaphor of the new age often mixes with the language of evolution, with technology adapting and improving inexorably over time. Combining these two elements produces a powerful sense of destiny or doom that has been widely taken up across media, industry, and government. Technologists in particular have internalized this view, describing themselves as being "caught in a tide, swept along in a current they cannot fight."[3] Technology will take a prescribed and unalterable course. It has already been decided; there is no alternative.

If the future is predetermined, then the same moral applies: lean forward into the future of work or be left behind in the dust of the past. Get onboard or get wrecked. Automation's ideal subject is the first to recognize this new reality. Of course, this version of automation comes attached with a bundle of technocratic norms and values. Gamification and incentivisation, virtuous cycles and venture capital, monetization and quantification—these are all hallmarks of a technologically driven approach to work. And these values sit in the broader dictates of capitalism, where the accumulation of profit and expansion of operations is axiomatic. And so the implied subject is also a docile subject, an acquiescent figure who affirms and even internalizes these values. After all, in the words of *Wired* editor Kevin Kelly, this is "what technology wants."[4]

Those who suggest otherwise are derided as technophobes, unable to get with the times. Or to use a favorite slur of technologists, they are neo-Luddites, today's versions of the English Luddites of the early nineteenth century who destroyed textile machinery as a form of protest. (Never mind that this irrational, reactive framing of the movement is a total mischaracterization.) Workers who wish for alternatives are inflexible or even ungrateful, wanting to move backward. As Crary notes, "to characterize current arrangements, in reality untenable and unsustainable, as anything but inevitable and unalterable is a contemporary heresy."[5] Voices who criticize this future of work and its invisible hand are blasphemous, failing to keep their faith in technically driven progress.

To push against the myth of automation, then, is to doubt the gospel of inevitability. It is to suggest, building on the previous chapters, that automation is not "our future" because that statement implies a singular future in which a singular humanity is impacted by a singular technology sweeping across the globe. Rather than a totalizing technology, Part 1 argued that these shifts are piecemeal and partial, reconfiguring labor rather than replacing humans. Rather than a singular future, Part 2 suggested there will be multiple futures. From nation to nation, region to region, and even city to city, technological development and adoption will play out in fundamentally different ways and at different speeds. And rather than a singular humanity, Part 3 demonstrated the racialized and gendered relations that elevate some humans while suppressing others. The all-embracing "our" is debunked by the long, tragic tale of history, showing that we are not all in this together. Automation's human costs are uneven, and this inequality means that we can only speak of "our futures" in relation to "their futures."

These counterpoints to automation's fated future are simple but fundamental. They take its closed future and fracture it into a constellation of open possibilities. Instead of a future-on-rails, we get futures-of-potential.

Indeed, as soon as we start to imagine in this space, we can see how monolithic the future of automation offered to us is. What's remarkable about the scenarios presented by industrial and tech firms like Siemens, IBM, Google, and others is just how similar they are. Whether presented under terms like "the future of automation" or "the future of health care," the ubiquitous screens, bleached pine, and manicured workspaces in these visions gesture to a kind of whitewashed world, a globally homogeneous realm where cultural, social, and racial difference has been flattened into oblivion. In these worlds, we are finally all on the same page. Politics, inequality, and antagonism of any type have been erased. Laborers seem to be doe-eyed automatons themselves, dutifully carrying out their tasks without a hint of protest or complaint. Work and life proceed with a smooth blandness. "Industry-led visions of the technological future have always been disingenuous" writes feminist futurist Sarah Kember, "we've never been more in need of alternatives."[6]

Alternative Values, Alternative Visions

Pushing back against automation as "our future," then, is a way to go beyond these industry-led visions and their unending sameness. These are more specific but also more radical dreams, where communities draw on their unique aspirations and capacities to imagine their own future of work. In these revised futures, the empire building, economic inequality, and environmental devastation that characterize contemporary technology may be replaced with more emancipatory visions or more ecological values.

For a glimpse of what these values might look like, we can turn briefly to the work of Mike Cooley and his coworkers at Lucas Aerospace. In the 1970s, these workers were facing structural unemployment due to the introduction of automated technologies. In response, they drafted the Lucas Plan, compiling a master list of the expertise and resources at their collective disposal. By consulting with community groups, polytechnics, and health providers, the Lucas shop stewards came up with concrete proposals for an alternative set of products they could manufacture—a mobility device for those with spina bifida, a life-support system, a road/rail vehicle, a kidney dialysis machine, and many others. In this flurry of activity, 150 designs were created. For some of the engineers, pouring their expertise into these designs for socially unmet needs became some of the most enriching projects of their lives.

The plan enabled those at the frontline of automation to assert their agency. Rather than allowing the future direction of technology to be passively dictated to them, they would actively shape its use and values. By mixing the hands-on knowledge of machinists with the theoretical proficiency of designers, workers could develop technologies that would better the lives of those around them. For Cooley, two clear pathways stood before them:

"Either we will have a future in which human beings are reduced to a sort of bee-like behavior, reacting to the systems and equipment specified for them; or we will have a future in which masses of people, conscious of their skills and abilities in both a political and a technical sense, decide that they are going to be the architects of a new form of technological development."[7] The question was simple: Would they be architects or bees?

In essence, the plan attempted to transition Lucas from a company focused on arms to one focused on the community. In comparison with the axioms of the military-industrial sector they usually served, this "socially useful production" would be driven by a very different set of priorities. While "socially useful" was never formally defined, Cooley derives a list of principles based on the kinds of projects that emerged from this push.[8] Formulated decades ago, these principles still resonate strongly—not least because they deftly blend strong humanistic and environmental values with operational guidelines drawn from a deep understanding of technical systems. Drawing closely on that work, I have condensed and updated his list to arrive at ten key principles for socially useful automation:

1. Automated work, from production to use and repair of a product, should be non-alienating.

2. Simple, safe, robust design, rather than complex brittle systems, should be championed.

3. Automated technologies should be visible and understandable to workers, with systems controlled by human beings rather than the reverse.

4. Automation should aim to minimize waste and conserve energy and materials.

5. Automation should support ecologically desirable and sustainable processes and products, from manufacture through to end use, repair, and disposal.

6. Automation should help and liberate human beings rather than constrain, control, or physically or mentally damage them.

7. Automation should assist minorities, disadvantaged groups, and those materially and otherwise deprived.

8. Automation should foster cooperation between producers and consumers, forge connections between nation-states, and enable mutually non-exploitative relationships between the developed and the developing world.

9. Automated processes and production should be regarded as part of culture, reflecting the cultural, historical, and other requirements of those who employ them.

10. Automation should not be concerned merely with production, but with the reproduction of knowledge and competence.

This list is hardly exhaustive. Indeed, to suggest any kind of definitive set of principles begins to veer dangerously close to the singular and all-encompassing version of automation critiqued across these pages. Any real-world program would need to liberally add, delete, or modify the list to suit their particular needs. But if these elements are provisional, they gesture to a key point: the development of technology is not predetermined. The singular path forward assumed by our current intersection of capital and technology is by no means preordained or inevitable. In fact, its key tenets, to echo Crary, are untenable and unsustainable. Other futures, with other agendas and other values, are possible.

How to Build the Future
In the 1970s, Cooley and the other Lucas workers grappled with what this future might look like, interviewing patients, engaging with their local community, sharing design ideas, and prototyping new forms of technical production. What might this future-building work look like today? It might look like Data for Black Lives, a movement that brings together coders, activists, and artists to respond to contemporary technologies and create measurable change. DBL has been highly vocal in its criticism of automated injustice, consistently raising the alarm about racialized technologies such as facial recognition, predictive policing, and algorithmic decision making. Yet

alongside this important role as a watchdog, they've also been constructing an alternative vision of work, holding roundtables on community organizing, political power, and the future of black wealth. Their prolific collection of lectures and seminars, posted over the past three years, becomes a kind of archive for the future, signaling some of the key desires and drives that would lie at the core of any Black-led program. Their titles alone—from "Abolition and Self-Determination" to "Data, Disaster, and Collective Power" and "Education, Justice, and Mathematics"—gesture to a project with fundamentally different priorities from the industry-led vision sketched above.[9] This is how alternative futures are born.

Alongside DBL we could also look to the Māori Data Sovereignty Network, a research program and community of scholars based in Aotearoa New Zealand. Automation, as the previous chapters should have made clear, is underpinned at all points by data. This makes access and ownership over data a key political question with major implications when it comes to labor. What information is collected? How is this stored and shared? And who generates profit from these activities? Given its key role in the future of work, the network advocates for Māori involvement in the governance of data repositories, supports the development of Māori data infrastructure and security systems, and promotes the development of sustainable Māori digital businesses and innovations.[10] By forging connections between indigenous scholars, government agencies, and university departments, the network seeks to build consensus around these critical issues. The network recognizes that data sovereignty is not simply an abstract matter that can be left to policymakers and technologists, but one crucial for the ongoing well-being of Māori both collectively and individually. If data is increasingly decisive, then this future starts by staking a claim to its production and circulation.

These two visions overlap at points while diverging at others. Each emerges from a colonial history in a certain part of the world. Each draws on a unique cosmology that explains the creation of humans, their relationship to others and the environment, and the role of work within society. Each is deeply shaped by particular values, from concepts such as abolition and Black liberation to the protection of *taonga* (treasure) in Māori culture. And each is embedded within a particular ecosystem of institutions and industry partners who bring with them matters raised within their particular commu- nities and their collective experiences. This unique mixture of social, cultural, and political factors comes together to produce two very different looking futures—and this is precisely the point. So-called global technologies and their supposedly universal values must be rejected. Applying these top-down "solutions" and labeling them "the future of work" will not suffice. Instead, any future will need to emerge organically, from the ground up, based on the specific capabilities and concerns of its community.

If these futures are disparate, so are the subjects that inhabit and catalyze it. Rather than the ideal neoliberal subjects of futurist scenarios, who smile and nod while the wonders of technology unfold around them, these futures are populated by a fuller gamut of flesh-and-blood inhabitants. Yes, there are certainly activists, labor organizers, and artists here. But beyond these well- known critical figures, I'm interested in how everyday individuals approach and situate themselves in relation to technically mediated labor. Instead of the binaries of technofetishistic or technophobic, these figures allow them- selves to adopt a more fine-grained stance. They may opt out of individual tracking but choose to pool their information into a "data commons." They might reject facial recognition while employing high-tech robotics on the assembly line. They could embrace robotic process automation yet maintain that teaching must occur face-to-face. In place of an all-or-nothing stance to "technology," these examples gesture to a more ambivalent starting point, which then oscillates freely among enthusiasm, apathy, and antipathy based on the context and the application. Different technologies of automation have different stakes and require different responses.

How is this response determined? When it comes to labor, trust in tech- nology cannot be assumed; it must be earned. For Erica Smiley, discussing automation does not mean rejecting it but being undeterred in asking the key question: "Who benefits and how do we vie for the rewards that come with this technology?"[11] Smiley's astute intervention activates a barrage of questions. What are the rewards (higher incomes, time off, less injury) that might accompany this technology? How will these rewards be distributed when it comes to race, class, and gender? What are the concrete advantages and disadvantages? How will social relations and power asymmetries at the workplace be transformed? And more pessimistically, what is the flip side of all of this, the negative by-products that may take the form of increased pressures or more articulated punishments? The answers to these questions determine whether an automation technology will benefit a select few or collectively enhance the well-being of workers in a meaningful way.

Toward Critical Automation

These insights start to suggest a rough agenda for critical automation research, a road map moving from the bleak present to more emancipatory futures. One provisional starting point, as this book has aimed to do, is to contest our current version of automation and its assumed supremacy. This fable is powerful but unhelpful, derailing a more articulated set of problematics.

In an immediate sense, this means breaking automation down to be more specific. What, where, and who are we talking about when we talk about automation? Narrowing down these frames of reference would help us to better grasp the particular people and things that are being transformed and the sociotechnical forces at work. Who exactly are "the workers," and what are their histories and backgrounds? Young migrants in Singapore will have a different set of precarities from working-class whites in Sydney, conditioning their ability to respond in resilient ways to technical upheavals. Bringing their lifeworld into the frame is vital for understanding how certain groups might best take hold of their futures in a self-determined way.

In a similar way, the "where" of automation cannot be left as some vague "wherever." Digging into particular instances of automated technologies quickly foregrounds the differences that exist from country to country, region to region, and even city to city. Technologies emerge from a place's unique ecosystem of institutions, innovations, and individuals, becoming cultural forms that reflect cultural values. Even supposedly "global" technologies get adapted as they touch down at the local level, domesticated for certain needs and markets. To speak of automated surveillance in the same breath as automated logistics is to skip over these local conditions, missing the deep geographical and cultural factors that could provide insights into their logics and drivers.

Along this focusing work, however, we also need to expand the conversation, to bring in more people from more fields with more viewpoints. Economics has thoroughly dominated much of the discussion around automation. Whether optimistic or pessimistic, its findings are based on quantifiable and often high-level statistics: average incomes and productivity curves and unemployment figures. These hard figures and their monetary view of "value" are a very particular way of understanding the world. But if automation is a problem, it is a wicked one, cutting across a number of disciplinary divides. This is a social, cultural, and political project as much as a technological one. A siloed approach where sociologists work in one corner, computer scientists in another, and race and cultural scholars in a third will not suffice. Any critical engagement with automation will need to overflow these tidy compartments, understanding technical logics but also folding in powerful insights from history and the humanities. This both/and approach would break new ground, developing a theorization that could take into account code and capital, operations and representations, processes and people.

By sharpening the focus and thickening the description, the grand sweep of "AUTOMATION" in all caps is reduced to a real research object that is at once messier and more interesting—something more fragmented, more humble, more context bound. As the vision loses its sheen, it becomes less blinding and easier to assess. Better questions can be asked; better insights can be generated. Only then can we make progress on the urgent task at hand: replacing the sagging, singular "future of work" with thousands of new stories of radical potential.

Acknowledgments

The catalyst for this book was the Geopolitics of Automation, a project funded by the Australian Research Council and led by Professors Brett Neilson and Ned Rossiter at the Institute for Culture and Society at Western Sydney University. Some material from this book comes from the early exploratory work I did for that project looking into automated technologies within Amazon and Alibaba warehouses. Many thanks to Ned and Brett for taking me on, giving me freedom to follow different threads, and providing insightful feedback.

Thanks is also due to the "Media, Race, Violence" reading group organized by Andrew Brooks at the School of Arts and Media at UNSW. This crew inspired a lot of the initial work around whiteness and automation, contributed readings around race and technology, and offered helpful feedback on writing.

Thanks to Erica Wetter at Stanford University Press for taking on the book and for her enthusiasm and expertise in shepherding it to publication.

An earlier and much shorter version of this essay appeared as a chapter in the book *Materializing Digital Futures: Touch, Movement, Sound and Vision* (Bloomsbury Academic, an imprint of Bloomsbury Publishing Inc, 2021). Thanks to editors Toija Cinque and Jordan Vincent at Deakin University for providing suggestions and prodding me to clarify certain points.

Of course, the greatest thanks goes to my family, Kimberlee, Ari, and Sol. My care for you and your futures inspires work like this; your care for me props me up and keeps me going.

Notes

Introduction

1. Harari, "How to Survive the 21st Century."
2. Schwartz, "Yes, the Robots Are Coming."
3. Michalski, "AI, Robotics, and the Future of Jobs."
4. Cupitt, *The World to Come*, 29.
5. Kelly, *The Inevitable*, jacket summary.
6. Mumford, *The Myth of the Machine*, 224.
7. Raghav, *Bullshit Jobs in China*, 9.
8. Easterling, *Enduring Innocence*, 103.
9. Beller, *The World Computer*, 10.

Chapter 1

1. Harper, "Automatic."
2. Susskind, *A World without Work*, 12.
3. Evans, *The Young Mill-Wright & Miller's Guide*.
4. ASME, "A. O. Smith Automatic Frame Plant."
5. Wesson, "Materials Handling."
6. Witt, "Automatic Storage and Retrieval System Control."
7. Elgozy, *Automation et humanisme*, 11.
8. Holusha, "General Motors: A Giant in Transition."
9. Brynjolfsson and McAfee, *Race against the Machine*, 23.
10. Ford, *Rise of the Robots*, 14.
11. Brynjolfsson and McAfee, *Race against the Machine*, 27.
12. Linder, "Are We Inching toward the Lights Out Factory?"

13. Debord, "Tesla's Future Is Completely Inhuman—and We Shouldn't Be Surprised."

14. Noble, *Forces of Production*; Braverman, *Labor and Monopoly Capital*.

15. Danaher, *Automation and Utopia*.

16. Srnicek and Williams, *Inventing the Future*.

17. Bastani, *Fully Automated Luxury Communism*.

18. Keynes, "Economic Possibilities for Our Grandchildren."

19. Stark, "Does Machine Displace Men in the Long Run?"

20. Vonnegut, *Player Piano*.

21. Time, "The Automation Jobless."

22. Pelegrin, "An Argument against Automation."

23. Barrett, "A Further Digression on the Over-Automated Warehouse."

24. Duan et al., "Balanced Order Batching with Task-Oriented Graph Clustering," 3.

25. Ray, "OpenAI's Gigantic GPT-3 Hints at the Limits of Language Models for AI."

26. Floridi, "AI and Its New Winter: From Myths to Realities."

27. Ingrassia and White, *Comeback*, 111.

28. Finkelstein, "GM and the Great Automation Solution."

29. Musk, "@timkhiggins Yes, Excessive Automation at Tesla Was a Mistake. To Be Precise, My Mistake. Humans Are Underrated."

30. Brynjolfsson and Beane, "Working with Robots in a Post-Pandemic World."

31. Smith, *Smart Machines and Service Work*, 11.

32. Russell, "Automated Picking Bins Set to Help New Zealand Growers Save Labour Costs."

33. Evans, "Not Even Wrong."

34. Benanav, "Automation and the Future of Work—2."

35. AlibabaTech, *Cainiao Smart Warehouse*.

36. Feng, "Couriers Quit before China's Ecommerce Shopping Festival, Demanding Better Pay and Work Conditions."

37. Simon, "Inside the Amazon Warehouse Where Humans and Machines Become One."

38. Seppelt and Lee, "Keeping the Driver in the Loop: Dynamic Feedback to Support Appropriate Use of Imperfect Vehicle Control Automation"; Endsley, "The Limits of Highly Autonomous Vehicles"; Kong et al., "Industrial Wearable System."

39. Haraway, *Simians, Cyborgs, and Women*.

40. Dewdney, *Last Flesh: Life in the Transhuman Era*.

41. TechCrunch, *Canvas' Robot Cart Could Change How Factories Work*.

42. Heater, "Amazon Acquires Autonomous Warehouse Robotics Startup Canvas Technology."

43. Mahroof, "A Human-Centric Perspective Exploring the Readiness towards Smart Warehousing."

44. Gue, "The Human-Centric Warehouse."

45. Marx, *Capital: A Critique of Political Economy*; Frey, "The Industrial Revolution and Its Discontents."

46. Taylor, *The Principles of Scientific Management*, 59.

47. Boguslaw, quoted in Cooley, *Architect or Bee?*, 41.

48. Mindell, *Our Robots, Ourselves: Robotics and the Myths of Autonomy*, 14.

49. Mindell, 15.

50. Simon, "Inside the Amazon Warehouse Where Humans and Machines Become One."

51. Heater, "These Are the Robots That Help You Get Your Amazon Packages on Time."

52. Amoore, *Cloud Ethics: Algorithms and the Attributes of Ourselves and Others*, 62.

53. Amoore, 64.

54. Crary, 24/7, 73.

55. Fannin, "The Rush to Deploy Robots in China amid the Coronavirus Outbreak."

56. Yongzhou, "Why Cainiao's Special and How Its Elastic Scheduling System Works."

57. Gui, "Crisis Is a Test of Our Spiritual Strength."

58. Amazon Jobs, "Results for Automation Engineer."

59. Amazon Jobs, "Automation Engineer in Kolbaskowo Poland."

60. Jünger, *The Failure of Technology*, 58.

61. Robinson, *New York 2140*, 406.

Chapter 2

1. Gray and Suri, *Ghost Work*.

2. Gray and Suri, 21.

3. Gray and Suri, 21.

4. International Labour Organization, "World Employment and Social Outlook," 143.

5. Jäger et al., "Crowdworking," 763.

6. Newlands and Lutz, "Crowdwork and the Mobile Underclass."

7. Dubal, "Digital Piecework."

8. Mateescu and Elish, "AI in Context," 36.

9. Mateescu and Elish, 12.

10. Merchant, "Why Self-Checkout Is and Has Always Been the Worst."

11. Merchant.

12. Hunt, "What Are Some Grocery Store Self-Checkout Hacks and Tricks?"

13. Taylor, "Supermarket Self-Checkouts and Retail Theft."

14. Mateescu and Elish, "AI in Context," 47.

15. Mateescu and Elish, 6.

16. Gribbin, "Japanese Knock Spots Off Unmanned Factories."

17. Raghav, *Bullshit Jobs in China*, 13.

18. TSC Foods, "Machine Minder."

19. Recruitme, "Machine Minder Rockit Packhouse."

20. Gillespie, *Custodians of the Internet*, 206.

21. Roberts, *Behind the Screen*, 139.

22. Roberts, Content Moderation of Social Media.

23. Mattes and Mattes, *Dark Content*.

24. Anonymous, "Resignation Note."

25. Lewig and McLean, "Caring for Our Frontline Child Protection Workforce."

26. NAMI, "Law Enforcement."

27. Newton, "Facebook Will Pay $52 Million in Settlement with Moderators Who Developed PTSD on the Job."

28. Roberts, Content Moderation of Social Media.

29. Roberts, *Behind the Screen*, 127.

30. Berardi, *After the Future*, 90.

31. Morris-Suzuki, "Robots and Capitalism," 113.

32. Morris-Suzuki, 114.

33. Munn, *Ferocious Logics*.

34. Goodwin, "The Battle Is for the Customer Interface."

35. Marx, *Capital*, 544.

36. Parisi, "Improper Commonness," 68.

37. Parisi, 74.

38. Parisi, 74.

39. Jäger et al., "Crowdworking."

40. van Doorn, "Platform Labor," 908.

41. van Doorn, 908.

42. Rivera, *Sleep Dealer*.

43. Teplitzky, "Remote Warfare, Automation, and Digital Labor."

44. Berardi, "Soul on Strike," 9.

Chapter 3

1. Ford, *Rise of the Robots*.

2. Hawksworth, Berriman, and Goel, "Will Robots Really Steal Our Jobs?"

3. Oppenheimer, *The Robots Are Coming!*

4. Rifkin, *The End of Work*.

5. Susskind, *A World without Work*.

6. Brynjolfsson and McAfee, *The Second Machine Age*.

7. Munn, "Red Territory: Forging Infrastructural Power."

8. Lawder and Heavey, "U.S. Blacklists China's Huawei as Trade Dispute Clouds Global Outlook."

9. Ernst, "Competing in Artificial Intelligence Chips: China's Challenge amid Technology War."

10. de Seta, "Into the Red Stack."

11. Hui, "Designing Tomorrow's Intelligence."

12. Yang, "A Chinese Internet?," 49.

13. Hui, *The Question Concerning Technology in China: An Essay in Cosmotechnics*.

14. Mosco, *The Digital Sublime*, 118.

15. Goodrich et al., "Semiconductors in the U.S.-China Tech Dispute."

16. Han, *Shanzhai: Deconstruction in Chinese*, 65.

17. Lindtner, Greenspan, and Li, "Designed in Shenzhen," 5.

18. De Greve, Zaman, and Schoffelen, "The Shanzhai City," 552.

19. Gu and Shea, "Fabbing the Chinese Maker Identity," 275.

20. Sampath and Khargonekar, "Socially Responsible Automation"; Wilshaw, "The Ethical Automation Toolkit."

21. Han, *Shanzhai: Deconstruction in Chinese*, 11.

22. Roberts et al., "The Chinese Approach to Artificial Intelligence."

23. Chien et al., "Influence of Culture, Transparency, Trust, and Degree of Automation on Automation Use."

24. Leung and Cohen, "Within- and Between-Culture Variation."

25. Anankaphannan et al., "Opportunities in the Chinese Logistics Industry," 10.

26. Falcone, Kent, and Fugate, "Supply Chain Technologies, Interorganizational Network and Firm Performance," 337.

27. Tsin, "Betting Big on Technification in South-East Asia."

28. Falcone, Kent, and Fugate, "Supply Chain Technologies, Interorganizational Network and Firm Performance," 337.

29. Falcone, Kent, and Fugate.

30. Tao and Shi, "Logistic Supply and Demand Analysis on Online Shopping of Double Eleven Day," 1457.

31. Rossiter, *Software, Infrastructure, Labor: A Media Theory of Logistical Nightmares*, 198.

32. Mezzadra and Neilson, "Extraction, Logistics, Finance," 15.

33. Kaplan, "Alibaba's 2020 Singles Day Breaks Record, Attracts Luxury Brands."

34. Last Mile Prophets, *Cainiao's New Post Office Network Cainiao Post*.

35. Lucas, Woodhouse, and Feng, "Alibaba Fights with Courier for Control of Customer Data."

36. Ye, "Data Spat between Home Delivery Majors Causes Huge Confusion."

37. Barbrook and Cameron, "The Californian Ideology."

38. Leetaru, "Is a Fragmented Internet Inevitable?"

39. Castro, McLaughlin, and Chivot, "Who Is Winning the AI Race," 1.

40. Li and Etchemendy, "We Need a National Vision for AI."

41. Ernst, "Competing in Artificial Intelligence Chips: China's Challenge amid Technology War."

42. Lee, *AI Superpowers: China, Silicon Valley, and the New World Order.*

43. Khalid, "A Dirty Word in the U.S., 'Automation' Is a Buzzword in China."

44. Khalid.

45. Kadir, Broberg, and Souza da Conceição, "Designing Human-Robot Collaborations in Industry 4.0: Explorative Case Studies," 605.

46. Pacey, *The Culture of Technology*, 3.

47. Spatharou, Hieronimus, and Jenkins, "Transforming Healthcare with AI: The Impact on the Workforce and Organizations."

48. Eubanks, *Automating Inequality.*

49. Bellandi and Di Tommaso, "The Case of Specialized Towns in Guangdong, China."

50. Zheng and Kuroda, "The Impact of Economic Policy on Industrial Specialization and Regional Concentration of China's High-Tech Industries."

Chapter 4

1. BBC News, "Who Are the Uighurs and Why Is the US Accusing China of Genocide?"

2. Associated Press, "China Cuts Uighur Births with IUDs, Abortion, Sterilization."

3. Khatchadourian, "Surviving the Crackdown in Xinjiang."

4. Harwell and Dou, "Huawei Tested AI Software That Could Recognize Uighur Minorities and Alert Police, Report Says."

5. IPVM, "Huawei/Megvii Uyghur Alarms."

6. Davidson, "Alibaba Offered Clients Facial Recognition to Identify Uighur People, Report Reveals"; Bhuiyan, "Major Camera Company Can Sort People by Race, Alert Police When It Spots Uighurs."

7. Wang et al., "Facial Feature Discovery for Ethnicity Recognition."

8. Long, "Xinjiang Cell Phone Users Forced to Register With Real Names."

9. Phillips, "China Orders Hundreds of Thousands of Private Cars to Have GPS Trackers Installed for Monitoring."

10. Quanguo, "陈全国：筑起反恐维稳的铜墙铁壁 确保新疆社会大局和谐稳定_央广网." [Build a copper wall and iron wall to fight terrorism and maintain stability to ensure the overall harmony and stability of Xinjiang's society].

11. Grauer, "Millions of Leaked Police Files Detail Suffocating Surveillance of China's Uyghur Minority."

12. Jünger, *The Failure of Technology*, 70.

13. Human Rights Watch, "China."

14. Mozur, "One Month, 500,000 Face Scans."

15. Wang, "China's Algorithms of Repression."

16. Wang; Yang, "Xinjiang Phone App Exposes How Chinese Police Monitor Uighur Muslims."

17. Xin, "The Police Region of Xinjiang."

18. Byler, "I Researched Uighur Society in China for 8 Years."

19. Roberts, *The War on the Uyghurs*, 84.

20. Kaltman, *Under the Heel of the Dragon*.

21. Roberts, *The War on the Uyghurs*, 437.

22. Leibold, "Surveillance in China's Xinjiang Region," 2.

23. Amoore, *Cloud Ethics*, 6.

24. Foucault, *Abnormal*.

25. Ramzy and Buckley, "'Absolutely No Mercy.'"

26. Dooley, "Inside China's Internment Camps."

27. China Law Translate, "Xinjiang Uyghur Autonomous Region Regulation on De-Extremification."

28. Lim, "China: Re-Engineering the Uighur."

29. Lim.

30. New China TV, *The Rise of Modern Farming in Xinjiang's Cotton Fields*.

31. New China TV.

32. New China TV.

33. New China TV.

34. New China TV.

35. New China TV.

36. China Daily, 陈柳兵, "Xinjiang Cotton by the Numbers."

37. China News Network, "H&M、耐克们，对新疆棉花一无所知⊠." [H&M and Nike don't know anything about Xinjiang cotton!].

38. Zenz, "Coercive Labor in Xinjiang," 3.

39. Zenz, 4.

40. Zenz, 5. https://xw.qq.com/cmsid/20191226A0DO0V00

41. Dao, "Xinjiang Cotton."

42. Dao.

43. 浮图, "河南女工去新疆摘棉花挣钱　几天后就后悔，坐棉花地痛哭." [QQ.com]. [A female worker from Henan went to Xinjiang to pick cotton to make money. A few days later, she regretted it and cried bitterly while sitting on the cotton field].

44. 浮图. [QQ.com].

45. Zenz, "Coercive Labor in Xinjiang," 9.

46. Zenz, 14.

47. Zenz, 16.

48. Zenz, 18.

49. Sudworth, "Inside China's Scheme to Transfer Uighurs into Work."

50. Sudworth.

51. Zenz, "Coercive Labor in Xinjiang," 15.

52. 新疆维吾尔自治区人力资源和社会保障厅, "关于做好拾花等季节性劳务工作的通知." [Department of Human Resources and Social Security of Xinjiang Uygur Autonomous Region], [Notice on doing a good job in seasonal labor services such as picking flowers].

53. Zenz, "Coercive Labor in Xinjiang," 12.

54. Sudworth, "China's 'Tainted' Cotton."

55. Khatchadourian, "Surviving the Crackdown in Xinjiang."

56. Zenz, "Coercive Labor in Xinjiang," 5.

Chapter 5

1. New York Times, "Man Devoured by His Machines."

2. Boyle, "Machines Are Laughing at Men."

3. Brickman and Lehrer, *Automation, Education, and Human Values.*

4. Leontief and Duchin, *The Future Impact of Automation on Workers.*

5. Simon, *The Shape of Automation for Men and Management,* 96.

6. Ellul, *The Technological Society,* 9.

7. Asimov, "The Evitable Conflict."

8. Fairman, *I, the Machine.*

9. Frey and Osborne, "The Future of Employment."

10. Atanasoski and Vora, *Surrogate Humanity,* 28.

11. Melamed, "Racial Capitalism," 77.

12. Rhee, *The Robotic Imaginary: The Human and the Price of Dehumanized Labor,* 3.

13. Alexander, *The Mantra of Efficiency,* 165.

14. Wajcman, "Automation."

15. Frey, "The Industrial Revolution and Its Discontents."

16. Marshall, "'Automation' Today and in 1662," 149.

17. Diebold, *Automation*; Joint Economic Committee, *Automation and Technological Change.*

18. Noble, *Forces of Production.*

19. Marx, *Capital*; Keynes, "Economic Possibilities for Our Grandchildren"; Braverman, *Labor and Monopoly Capital.*

20. Greenfield, *Radical Technologies: The Design of Everyday Life,* 404.

21. Atanasoski and Vora, *Surrogate Humanity*, 4.

22. Robinson, *Black Marxism*, 2.

23. Moore, "Sugar and the Expansion of the Early Modern World-Economy," 189.

24. Moore, *Capitalism in the Web of Life: Ecology and the Accumulation of Capital*, 60.

25. Roediger, *The Wages of Whiteness*.

26. Morrison, *Playing in the Dark: Whiteness and the Literary Imagination*, 61.

27. Moradi, "Race, Ethnicity, and the Future of Work," 8.

28. Ding, "Study of Smart Warehouse Management System Based on the IOT."

29. Liu et al., "CPS-Based Smart Warehouse for Industry 4.0."

30. Hatt, "Racializing Smartness."

31. Wynter, "Unsettling the Coloniality of Being/Power/Truth/Freedom," 296.

32. Hilliard, *Straightening the Bell Curve*.

33. Plous and Williams, "Racial Stereotypes from the Days of American Slavery: A Continuing Legacy."

34. Sparrow, "Robotics Has a Race Problem"; Bartneck et al., "Robots and Racism."

35. Moore-Colyer, "Amazon's Human Workers Are Safe for Now, But the Tide of Automation Is Rising."

36. Fredrickson, *White Supremacy*, 21.

37. Melamed, "Racial Capitalism," 77.

38. Oliver and Shapiro, *Black Wealth/White Wealth*, 2.

39. Oliver and Shapiro, 53.

40. Price, "Black People vs Robots."

41. Cook, "The Future of Work in Black America."

42. Ahlen, "Impact of Robots on the Workforce."

43. Duffy, "Doing the Dirty Work: Gender, Race, and Reproductive Labor in Historical Perspective."

44. Glenn, "From Servitude to Service Work."

45. Price, "Black People vs Robots."

46. Anonymous, "Blue Collar Blues."

47. Pellow and Park, *The Silicon Valley of Dreams*, 13.

48. Amazon, "Our Workforce Data."

49. Demmitt, "85 Percent of Amazon's Black U.S. Workers Hold Unskilled Jobs."

50. Gutelius and Theodore, "The Future of Warehouse Work: Technological Change in the U.S. Logistics Industry."

51. Allen, "An Order Every 33 Seconds"; O'Connor, "Amazon Unpacked."

52. Scheiber, "Inside an Amazon Warehouse, Robots' Ways Rub Off on Humans."

53. Del Rey, "How Robots Are Transforming Amazon Warehouse Jobs—for Better and Worse."

54. Evans, "Ruthless Quotas at Amazon Are Maiming Employees"; Evans, "How Amazon Hid Its Safety Crisis."

55. Evans, "How Amazon Hid Its Safety Crisis."

56. Bezos, "2020 Letter to Shareholders."

57. Bezos.

58. Bruder, "Meet the Immigrants Who Took On Amazon."

59. Chen and Crabapple, "Meet the Warehouse Worker Who Took On Amazon over Inhumane Conditions and Harassment."

60. Bruder, "Meet the Immigrants Who Took On Amazon." For a parallel example of labor activism in China, we could look at Foxconn, where employees have walked off the job and engaged in rioting, shutting down entire assembly lines for that period and interrupting the production of the iPhone. While less obvious, race is a factor here too, with rural Chinese workers feeling they are looked down on by "superior" Taiwanese management. One worker wrote, "Workers live at the lowest level, Tolerating the most intense work, Earning the lowest pay, Accepting the strictest regulation, And enduring discrimination everywhere. Even though you are my boss, and I am a worker: I have the right to speak to you on an equal footing." All these points and quotes come from the excellent chapter by Chan, Pun, and Selden, "Chinese Labor Protest and Trade Unions."

61. Harney and Moten, *The Undercommons*, 67.

62. Andrejevic, *Automated Media*, 2.

63. Chen and Crabapple, "Meet the Warehouse Worker Who Took On Amazon Over Inhumane Conditions and Harassment."

64. Hopkins, Amazon Is No Ally in the Fight for Racial Justice.

65. Blest, "Leaked Amazon Memo Details Plan to Smear Fired Warehouse Organizer."

66. Carney, "Fired Amazon Designer."

67. Bray, "Bye, Amazon."

68. Wynter, "Unsettling the Coloniality of Being/Power/Truth/Freedom," 323.

69. Brémond, "Racism in Big Tech."

70. Turner-Lee, "Black People vs Robots."

71. Marx, *Capital*.

72. Scholz, "Platform Cooperativism."

73. Smiley, "Black People vs Robots."

74. Delaney, "The Robot That Takes Your Job Should Pay Taxes, Says Bill Gates."

75. Estlund, "What Should We Do after Work: Automation and Employment," 316.

76. Rushkoff, "Silicon Valley's Push for Universal Basic Income"; Sadowski, "Why Silicon Valley Is Embracing Universal Basic Income."

77. Price, "Black People vs Robots."

78. There is a long history of full personhood depending on being productive. As Sarah Rose shows so powerfully, the classification of the "disabled" historically was complex not just for society as a whole, but for policymakers, reformers, and government. There was a broad constellation of bodily infirmities across the population, from injured limbs to cognitive impairments. And how these were attained—during birth, from the "sins of the parents," or fighting for the country—complicated questions of blame and aid even further. In the face of this complexity, the question of labor became a powerful overarching filter: whether one could work a full day and "be useful." As Rose shows, families of "idiots" or the "feeble-minded" never fully subscribed to this black and white binary of productive and unproductive, integrating these individuals into work and home life where they contributed by fetching water, caring for children, tending livestock, and so on. And yet with the encroachment of industrialization in the early twentieth century (Rose's study focuses on the United States), the locations and definitions of work shifted. Employers' stance to those with disabilities hardened, not least because of the new risks associated with well-meaning legislation like workmen's compensation. Work became defined by efficiency, and the most efficient work could only be carried out by those who had "intact, interchangeable bodies." In this sense, mechanization is not only about achieving optimal workflow, but about standardizing the body of the worker herself. Standardization accepts particular kinds of bodies while filtering out those who do not fit its regimented, routinized approach. Those with differing bodies, as Rose demonstrates, were seen as drains on the system, "deadweights" who contributed nothing to society. Rose, *No Right to Be Idle*.

79. Smiley, "Black People vs Robots."

Chapter 6

1. Hester, "Technically Female."

2. Roberts et al., "The Future Is Ours: Women, Automation and Equality," 3.

3. Hegewisch, Childers, and Hartmann, "Women, Automation and the Future of Work."

4. Brussevich, Dabla-Norris, and Khalid, *Is Technology Widening the Gender Gap?*

5. Cortes and Pan, "Gender, Occupational Segregation, and Automation," 19.

6. Plant, *Zeroes + Ones: Digital Women + the New Technoculture*, 38.

7. Plant, 43.

8. Wells, "Automation and Women Workers."

9. Hegewisch, Childers, and Hartmann, "Women, Automation and the Future of Work."

10. Grierson, "Amazon 'Regime' Making British Staff Physically and Mentally Ill, Says Union."

11. Guendelsberger, *On the Clock*, 96.

12. Guendelsberger, 20.

13. weepinwidow, "Amazon and Mental Health."

14. HelloMya, "I Had My First Amazonian Mental Breakdown at Work Last Night."

15. Zahn and Paget, "'Colony of Hell.'"

16. Lieber, "Emergency Calls Placed from Amazon Warehouses Depict Enormous Pressure Put on Workers."

17. Lecher, "How Amazon Automatically Tracks and Fires Warehouse Workers for 'Productivity.'"

18. Noble, *Forces of Production*, 41.

19. Rowanissupreme, "The Mental Health at Amazon Seems Debilitating."

20. McKinsey Global Institute, "Automation, Jobs, and the Future of Work."

21. Brynjolfson and McAfee, "The Second Machine Age," 35.

22. Ransome, *Work, Consumption and Culture*, 24.

23. Ferguson, *Women and Work*, 120.

24. Hester, "Technically Female."

25. Federici, "The Reproduction of Labour Power in the Global Economy and the Unfinished Feminist Revolution," 88.

26. Federici, 89.

27. Bassett, Kember, and O'Riordan, *Furious: Technological Feminism and Digital Futures*, 46.

28. Wajcman, *Pressed for Time: The Acceleration of Life in Digital Capitalism*, 122.

29. Cowan, *More Work for Mother*.

30. Cowan, 210.

31. Smith, *Smart Machines and Service Work*, 130.

32. Rothschild, *Women, Technology, and Innovation*; Zimmerman, *The Technological Woman*.

33. SystematicApproach, "Wife Sucks the Fun out of HA."

34. Berg, "The Smart House as a Gendered Socio-Technical Construction."

35. Chambers, *Changing Media, Homes and Households*, 156.

36. Turk, "Home Invasion."

37. Schwedel, "When Alexa Becomes Part of the Family."

38. Purington et al., "Alexa Is My New BFF: Social Roles, User Satisfaction, and Personification of the Amazon Echo."

39. McGirt, "Amazon's Echo Device Chief."

40. Han, *Psychopolitics*, 36.

41. Gold, "Fembots Have Feelings Too."

42. Schwartz, "Untold History of AI."

43. Schwartz.

44. Hicks, *Programmed Inequality*, 105.

45. Hicks, 224.

46. Hicks, 224.

47. Hicks, 420.

48. Hicks, 174.

49. Hicks, 174.

50. Hicks, 426.

51. Bassett, Kember, and O'Riordan, *Furious: Technological Feminism and Digital Futures*, 67.

52. McAlevey, with Ostertag, *Raising Expectations (and Raising Hell)*, 30.

53. Hester and Srnicek, "The Crisis of Social Reproduction and the End of Work."

Conclusion

1. Brynjolfsson and McAfee, *The Second Machine Age*, 229.

2. Crary, *24/7*, 36.

3. Eveleth, "The Biggest Lie Tech People Tell Themselves—and the Rest of Us."

4. Kelly, *What Technology Wants*.

5. Crary, *24/7*, 39.

6. Kember, "Notes Towards a Feminist Futurist Manifesto."

7. Cooley, *Architect or Bee?*, 180.

8. Cooley, 155.

9. Data for Black Lives, "Data for Black Lives Videos."

10. Te Mana Raraunga, "What We Do."

11. Smiley, "Black People vs Robots."

References

Ahlen, J. E. "Impact of Robots on the Workforce." California State University, 1987. https://www.elibrary.ru/item.asp?id=6758431.

Alexander, Jennifer Karns. *The Mantra of Efficiency: From Waterwheel to Social Control*. Baltimore: Johns Hopkins University Press, 2008.

AlibabaTech. *Cainiao Smart Warehouse*. 2018. https://www.youtube.com/watch?v=jRYaAGH_Z-0.

Allen, Vanessa. "An Order Every 33 Seconds: How Amazon Works Staff 'to the Bone.'" *Mail Online*, November 25, 2013. http://www.dailymail.co.uk/news/article-2512959/Walk-11-miles-shift-pick-order-33-seconds--Amazon-works-staff-bone.html.

Amazon. "Our Workforce Data." *About Amazon*, April 10, 2019. https://www.aboutamazon.com/news/workplace/our-workforce-data.

Amazon Jobs. "Automation Engineer in Kolbaskowo Poland." Amazon Jobs, 2021. https://www.amazon.jobs/en/jobs/1280897/automation-engineer.

———. "Results for Automation Engineer." Amazon Jobs, 2021. http://www.amazon.jobs/en/search?base_query=automation+engineer&loc_query=&latitude=&longitude=&loc_group_id=&invalid_location=false&country=&city=&region=&county=.

Amoore, Louise. *Cloud Ethics: Algorithms and the Attributes of Ourselves and Others*. Durham, NC: Duke University Press, 2020.

Anankaphannan, Gift, Wenting Want, Eric Liu, Hardy Zhang, Terrance Fang, and Denise Cai. "Opportunities in the Chinese Logistics Industry." Fontainebleau, France: INSEAD, May 2012. https://www.insead.edu/sites/default/files/assets/dept/centres/gpei/docs/insead-student-opportunities-in-the-chinese-logistics-industry-may-2012.pdf.

Andrejevic, Mark. *Automated Media*. London: Routledge, 2019.

Anonymous. "Blue Collar Blues." *Digger*, March 1974. https://issuu.com/libuow/docs/digger1974n028_cc68e788ee7459.

———. "Resignation Note," April 14, 2021. https://pastebin.com/raw/12e4YaX7.

ASME. "A. O. Smith Automatic Frame Plant." New York: American Society of Mechanical Engineers, 2020. https://www.asme.org/about-asme/engineering-history/landmarks/37-a-o-smith-automatic-frame-plant.

Asimov, Isaac. "The Evitable Conflict." *Astounding Science Fiction* 45, no. 4 (1950): 48–68.

Associated Press. "China Cuts Uighur Births with IUDs, Abortion, Sterilization." *AP News*, June 29, 2020. https://apnews.com/article/269b3de1af34e17c1941a514f78d764c.

Atanasoski, Neda, and Kalindi Vora. *Surrogate Humanity: Race, Robots, and the Politics of Technological Futures*. Durham: Duke University Press, 2019.

Barbrook, Richard, and Andy Cameron. "The Californian Ideology." *Metamute* 1, no. 3 (1995). https://www.metamute.org/editorial/articles/californian-ideology.

Barrett, Bruce G. "A Further Digression on the Over-Automated Warehouse: Some Evidence." *Interfaces* 8, no. 1 (1977): 46–49.

Bartneck, Christoph, Kumar Yogeeswaran, Qi Min Ser, Graeme Woodward, Robert Sparrow, Siheng Wang, and Friederike Eyssel. "Robots and Racism." In *Proceedings of the 2018 ACM/IEEE International Conference on Human-Robot Interaction*, 196–204. Piscataway, NJ: IEEE, 2018.

Bassett, Caroline, Sarah Kember, and Kate O'Riordan. *Furious: Technological Feminism and Digital Futures*. London: Pluto Press, 2019.

Bastani, Aaron. *Fully Automated Luxury Communism*. London: Verso Books, 2019.

BBC News. "Who Are the Uighurs and Why Is the US Accusing China of Genocide?" *BBC News*, February 9, 2021. https://www.bbc.com/news/world-asia-china-22278037.

Bellandi, Marco, and Marco R. Di Tommaso. "The Case of Specialized Towns in Guangdong, China." *European Planning Studies* 13, no. 5 (2005): 707–29. https://doi.org/10.1080/09654310500139244.

Beller, Jonathan. *The World Computer: Derivative Conditions of Racial Capitalism*. Durham: Duke University Press, 2021.

Berardi, Franco. *After the Future*. Translated by Arianna Bove. Edinburgh: AK Press, 2011.

———. "Soul on Strike." In *The Soul at Work*, 9–20. Cambridge, MA: MIT Press, 2009. https://mitpress.mit.edu/books/soul-work.

Berg, Anne-Jorunn. "The Smart House as a Gendered Socio-Technical Construction." Working Paper 14/92. Trondheim, Norway: Centre for Technology and Society, 1992.

Bezos, Jeff. "2020 Letter to Shareholders." *About Amazon*, April 15, 2021. https://www.aboutamazon.com/news/company-news/2020-letter-to-shareholders.

Bhuiyan, Johana. "Major Camera Company Can Sort People by Race, Alert Police When It Spots Uighurs." *Los Angeles Times*, February 9, 2021. https://www.latimes.com/business/technology/story/2021-02-09/dahua-facial-recognition-china-surveillance-uighur.

Blest, Paul. "Leaked Amazon Memo Details Plan to Smear Fired Warehouse Organizer: 'He's Not Smart or Articulate.'" *Vice News*, March 4, 2020. https://www.vice.com/en/article/5dm8bx/leaked-amazon-memo-details-plan-to-smear-fired-warehouse-organizer-hes-not-smart-or-articulate.

Boggs, James. *The American Revolution: Pages from a Negro Worker's Notebook*. New York: Monthly Review Press, 1968.

Boyle, Hal. "Machines Are Laughing at Men." *Evening Independent*, April 14, 1949. https://news.google.com/newspapers?nid=950&dat=19490414&id=LPhPAAAAIBAJ&sjid=glUDAAAAIBAJ&pg=5992,5416204&hl=en.

Braverman, Harry. *Labor and Monopoly Capital: The Degradation of Work in the Twentieth Century*. New York: Monthly Review Press, 1998.

Bray, Tim. "Bye, Amazon." *Ongoing*, April 29, 2020. https://www.tbray.org/ongoing/When/202x/2020/04/29/Leaving-Amazon#p-1.

Brémond, Anaïs. "Racism in Big Tech: Amazon Workers Speak Up." *Welcome to the Jungle*, July 8, 2020. https://www.welcometothejungle.com/en/articles/racism-big-tech-amazon-warehouse-workers-speak.

Brickman, William W., and Stanley Lehrer. *Automation, Education, and Human Values*. New York: School & Society Books, 1966.

Bruder, Jessica. "Meet the Immigrants Who Took On Amazon." *Wired*, November 12, 2019. https://www.wired.com/story/meet-the-immigrants-who-took-on-amazon/.

Brussevich, Mariya, Era Dabla-Norris, and Salma Khalid. *Is Technology Widening the Gender Gap? Automation and the Future of Female Employment*. Washington, DC: International Monetary Fund, 2019.

Brynjolfsson, Erik, and Matt Beane. "Working with Robots in a Post-Pandemic World." *MIT Sloan Management Review*, September 16, 2020. https://sloanreview.mit.edu/article/working-with-robots-in-a-post-pandemic-world/.

Brynjolfsson, Erik, and Andrew McAfee. *Race against the Machine: How the Digital Revolution Is Accelerating Innovation, Driving Productivity, and Irreversibly Transforming Employment and the Economy*. Boston: Digital Frontier Press, 2011.

———. *The Second Machine Age: Work, Progress, and Prosperity in a Time of Brilliant Technologies*. New York: Norton, 2014.

Byler, Darren. "I Researched Uighur Society in China for 8 Years and Watched How Technology Opened New Opportunities—Then Became a Trap." *Conversation*, September 18, 2019. http://theconversation.com/i-researched-uighur-society-in-china-for-8-years-and-watched-how-technology-opened-new-opportunities-then-became-a-trap-119615.

Carney, Bryan. "Fired Amazon Designer: I Was Punished for Uniting Tech and Warehouse Workers." *Tyee*, May 18, 2020. https://thetyee.ca/News/2020/05/18/Fired-Amazon-Designer-Punished/.

Castro, Daniel, Michael McLaughlin, and Eline Chivot. "Who Is Winning the AI Race: China, the EU or the United States?" Washington, DC: Center for Data Innovation, August 19, 2019. https://datainnovation.org/2019/08/who-is-winning-the-ai-race-china-the-eu-or-the-united-states/.

Chambers, Deborah. *Changing Media, Homes and Households: Cultures, Technologies and Meanings.* London: Routledge, 2016.

Chan, Jenny, Ngai Pun, and Mark Selden. "Chinese Labor Protest and Trade Unions." In *The Routledge Companion to Labor and Media*, edited by Richard Maxwell, 290–302. New York: Routledge, 2015.

Chen, Michelle, and Molly Crabapple. "Meet the Warehouse Worker Who Took On Amazon over Inhumane Conditions and Harassment." *In These Times*, October 5, 2020. https://inthesetimes.com/article/amazon-profile-hibaq-mohammed.

Chien, S., M. Lewis, K. Sycara, A. Kumru, and J. Liu. "Influence of Culture, Transparency, Trust, and Degree of Automation on Automation Use." *IEEE Transactions on Human-Machine Systems* 50, no. 3 (2020): 205–14. https://doi.org/10.1109/THMS.2019.2931755.

陈柳兵. [*China Daily.*] "Xinjiang Cotton by the Numbers." March 27, 2021. https://www.chinadaily.com.cn/a/202103/27/WS605e6b94a31024adobab1f12.html.

China Law Translate. "Xinjiang Uyghur Autonomous Region Regulation on De-Extremification." March 30, 2017. https://www.chinalawtranslate.com/xinjiang-uyghur-autonomous-region-regulation-on-de-extremification/.

China News Network. "H&M、耐克们，对新疆棉花一无所知！，" [H&M and Nike don't know anything about Xinjiang cotton!]. March 25, 2021. http://www.chinanews.com/gn/2021/03-25/9439972.shtml.

Cinque, Toija, and Jordan Beth Vincent, eds. *Materializing Digital Futures: Touch, Movement, Sound and Vision.* New York: Bloomsbury Academic, 2021. https://www.bloomsbury.com/us/materializing-digital-futures-9781501361265.

Clark, Charles. "Trade Unions Say That Amazon's 'Regime' Is Causing Its UK Employees to Suffer Mental and Physical Illness." *Business Insider*, August 18, 2015. https://www.businessinsider.com/amazons-regime-is-causing-employees-to-suffer-mental-and-physical-illness-2015-8.

Cook, Kelemwork. "The Future of Work in Black America." Washington, DC: McKinsey & Company, October 4, 2019. https://www.mckinsey.com/featured-insights/future-of-work/the-future-of-work-in-black-america#.

Cooley, Mike. *Architect or Bee? The Human Price of Technology*. Slough: Langley Technical Services, 1980.

Cortes, Patricia, and Jessica Pan. "Gender, Occupational Segregation, and Automation." Washington, DC: Brookings Institute, November 2019. https://www.brookings.edu/wp-content/uploads/2019/11/Cortes_Pan_Gender-occupational-segregation-and-automation.pdf.

Cowan, Ruth Schwartz. *More Work for Mother: The Ironies of Household Technology from the Open Hearth to the Microwave*. New York: Basic Books, 1985.

Crary, Jonathan. *24/7: Late Capitalism and the Ends of Sleep*. London: Verso, 2014.

Cupitt, Don. *The World to Come*. London: SCM Press, 1982.

Danaher, John. *Automation and Utopia: Human Flourishing in a World without Work*. Cambridge, MA: Harvard University Press, 2019.

Dao, Chang. "Xinjiang Cotton: Why Hand-Picking Is Still in Demand." Translated by New York MOS Translation Team RD16. *GNEWS*, March 28, 2021. https://gnews.org/1032892.

Data for Black Lives. "Data for Black Lives Videos." YouTube, 2021. https://www.youtube.com/c/DataforBlackLives/videos.

Davidson, Helen. "Alibaba Offered Clients Facial Recognition to Identify Uighur People, Report Reveals." *Guardian*, December 17, 2020. http://www.theguardian.com/business/2020/dec/17/alibaba-offered-clients-facial-recognition-to-identify-uighur-people.

Debord, Matthew. "Tesla's Future Is Completely Inhuman—and We Shouldn't Be Surprised." *Business Insider*, May 21, 2017. https://www.businessinsider.com.au/tesla-completely-inhuman-automated-factory-2017-5?r=UK.

De Greve, Teis, Bieke Zaman, and Jessica Schoffelen. "The Shanzhai City." In *Proceedings of the 31st Australian Conference on Human-Computer-Interaction*, 550–54. Fremantle: ACM, 2019.

Del Rey, Jason. "How Robots Are Transforming Amazon Warehouse Jobs—for Better and Worse." *Vox*, December 11, 2019. https://www.vox.com/recode/2019/12/11/20982652/robots-amazon-warehouse-jobs-automation.

Delaney, Kevin. "The Robot That Takes Your Job Should Pay Taxes, Says Bill Gates." *Quartz*, February 18, 2017. https://qz.com/911968/bill-gates-the-robot-that-takes-your-job-should-pay-taxes/.

Demmitt, Jacob. "85 Percent of Amazon's Black U.S. Workers Hold Unskilled Jobs." *Puget Sound Business Journal*, June 11, 2015. https://www.bizjournals.com/seattle/blog/techflash/2015/06/85-percent-of-amazon-s-black-u-s-workers-hold.html.

新疆维吾尔自治区人力资源和社会保障厅. "关于做好拾花等季节性劳务工作的通知." 商业新知网, [Department of Human Resources and Social Security of Xinjiang Uygur Autonomous Region]. [Notice on doing a good job in seasonal labor services such as picking flowers]. September 1, 2016. https://www.shangyexinzhi.com/article/1656034.html.

Dewdney, Christopher. *Last Flesh: Life in the Transhuman Era*. San Francisco: Harper, 1998.

Diebold, John. *Automation: The Advent of the Automatic Factory*. Princeton, NJ: Van Nostrand, 1952.

Ding, Wen. "Study of Smart Warehouse Management System Based on the IOT." In *Intelligence Computation and Evolutionary Computation*, 203–207. New York: Springer, 2013.

Dooley, Ben. "Inside China's Internment Camps: Tear Gas, Tasers and Textbooks." AFP, January 30, 2019. https://www.afp.com/en/inside-chinas-internment-camps-tear-gas-tasers-and-textbooks.

Doorn, Niels van. "Platform Labor: On the Gendered and Racialized Exploitation of Low-Income Service Work in the 'On-Demand' Economy." *Information, Communication and Society* 20, no. 6 (2017): 898–914. https://doi.org/10.1080/1369118X.2017.1294194.

Duan, Lu, Haoyuan Hu, Zili Wu, Guozheng Li, Xinhang Zhang, Yu Gong, and Yinghui Xu. "Balanced Order Batching with Task-Oriented Graph Clustering." In *Proceedings of the 26th ACM SIGKDD International Conference on Knowledge Discovery and Data Mining*, 3044–53. New York: ACM, August 23, 2020. https://doi.org/10.1145/3394486.3403355.

Dubal, Veena. "Digital Piecework." *Dissent Magazine*, Fall 2020. https://www.dissentmagazine.org/article/digital-piecework.

Duffy, Mignon. "Doing the Dirty Work: Gender, Race, and Reproductive Labor in Historical Perspective." *Gender & Society* 21, no. 3 (2007): 313–36.

Easterling, Keller. *Enduring Innocence: Global Architecture and Its Political Masquerades*. Cambridge, MA: MIT Press, 2005.

Elgozy, Georges. *Automation et humanisme*. Paris: Calmann-Lévy, 1968.

Ellul, Jacques. *The Technological Society*. London: Vintage, 1967.

Endsley, Mica R. "The Limits of Highly Autonomous Vehicles: An Uncertain Future." *Ergonomics* 62, no. 4 (2019): 496–99. https://doi.org/10.1080/00140139.2019.1563330.

Ernst, Dieter. "Competing in Artificial Intelligence Chips: China's Challenge amid Technology War." Waterloo, Canada: Centre for International Governance Innovation, 2020.

Estlund, Cynthia. "What Should We Do after Work: Automation and Employment." *Yale Law Journal* 128, no. 2 (2018): 254–326.

Eubanks, Virginia. *Automating Inequality: How High-Tech Tools Profile, Police, and Punish the Poor*. New York: St. Martin's Press, 2018.

Evans, Benedict. "Not Even Wrong: Predicting Tech." Benedict Evans, May 16, 2020. https://www.ben-evans.com/benedictevans/2020/5/16/not-even-wrong.

Evans, Oliver. *The Young Mill-Wright & Miller's Guide*. Philadelphia: Carey, Lea, and Blanchard, 1795. http://archive.org/details/youngmillwrightmooevan.

Evans, Will. "How Amazon Hid Its Safety Crisis." *Reveal*, September 29, 2020. https://revealnews.org/article/how-amazon-hid-its-safety-crisis/.

Evans, Will. "Ruthless Quotas at Amazon Are Maiming Employees." *Atlantic*, December 5, 2019. https://www.theatlantic.com/technology/archive/2019/11/amazon-warehouse-reports-show-worker-injuries/602530/.

Eveleth, Rose. "The Biggest Lie Tech People Tell Themselves—and the Rest of Us." *Vox*, October 1, 2019. https://www.vox.com/the-highlight/2019/10/1/20887003/tech-technology-evolution-natural-inevitable-ethics.

Fairman, Paul. *I, the Machine*. Abingdon, UK: Lodestone Books, 1968.

Falcone, Ellie, John Kent, and Brian Fugate. "Supply Chain Technologies, Interorganizational Network and Firm Performance: A Case Study of Alibaba Group and Cainiao." *International Journal of Physical Distribution and Logistics Management* 50, no. 3 (2019): 333–54. https://doi.org/10.1108/IJPDLM-08-2018-0306.

Fannin, Rebecca. "The Rush to Deploy Robots in China amid the Coronavirus Outbreak." CNBC, March 2, 2020. https://www.cnbc.com/2020/03/02/the-rush-to-deploy-robots-in-china-amid-the-coronavirus-outbreak.html.

Federici, Silvia. "The Reproduction of Labour Power in the Global Economy and the Unfinished Feminist Revolution." In *Workers and Labour in a Globalised Capitalism: Contemporary Themes and Theoretical Issues*, edited by Maurizio Atzeni, 85–107. London: Palgrave Macmillan, 2013.

Feng, Jiayun. "Couriers Quit before China's Ecommerce Shopping Festival, Demanding Better Pay and Work Conditions." *SupChina*, October 30, 2020. https://supchina.com/2020/10/30/couriers-quit-before-chinas-ecommerce-shopping-festival-demanding-better-pay-and-work-conditions/.

Ferguson, Susan J. *Women and Work: Feminism, Labour, and Social Reproduction*. London: Pluto Press, 2020.

Finkelstein, Sydney. "GM and the Great Automation Solution." *Business Strategy Review* 14, no. 3 (2003): 18–24. https://doi.org/10.1111/1467-8616.00268.

Floridi, Luciano. "AI and Its New Winter: From Myths to Realities." *Philosophy and Technology* 33 (2020): 1–3.

Ford, Martin. *Rise of the Robots: Technology and the Threat of a Jobless Future*. New York: Basic Books, 2016.

Foucault, Michel. *Abnormal: Lectures at the Collège de France, 1974–1975*. Translated by Graham Burchell. New York: Picador, 2003.

Fredrickson, George M. *White Supremacy: A Comparative Study of American and South African History*. Cambridge: Oxford University Press, 1981.

Frey, Carl Benedikt. "The Industrial Revolution and Its Discontents." In *The Technol-*

ogy Trap: Capital, Labor, and Power in the Age of Automation, 235–290. Princeton, NJ: Princeton University Press, 2020.

Frey, Carl Benedikt, and Michael A. Osborne. "The Future of Employment: How Susceptible Are Jobs to Computerisation?" *Technological Forecasting and Social Change* 114 (2013): 254–80. https://doi.org/10.1016/j.techfore.2016.08.019.

Gillespie, Tarleton. *Custodians of the Internet: Platforms, Content Moderation, and the Hidden Decisions That Shape Social Media*. New Haven: Yale University Press, 2018.

Glenn, Evelyn Nakano. "From Servitude to Service Work: Historical Continuities in the Racial Division of Paid Reproductive Labor." *Signs: Journal of Women in Culture and Society* 18, no. 1 (1992). https://doi.org/10.1086/494777.

Gold, Hannah. "Fembots Have Feelings Too." *New Republic*, May 12, 2015. https://newrepublic.com/article/121766/ex-machina-critiques-ways-we-exploit-female-care.

Goodrich, Jimmy, Derek Scissors, Willy Shih, and Clete Willems. "Semiconductors in the U.S.-China Tech Dispute." Presented at the Asia Society Northern California, November 10, 2020. https://www.youtube.com/watch?v=qs1PGD6AXPk.

Goodwin, Tom. "The Battle Is for the Customer Interface." *TechCrunch* (blog), March 3, 2015. http://social.techcrunch.com/2015/03/03/in-the-age-of-disintermediation-the-battle-is-all-for-the-customer-interface/.

Grauer, Yael. "Millions of Leaked Police Files Detail Suffocating Surveillance of China's Uyghur Minority." *Intercept*. Accessed February 15, 2021, at https://theintercept.com/2021/01/29/china-uyghur-muslim-surveillance-police/.

Gray, Mary L., and Siddharth Suri. *Ghost Work: How to Stop Silicon Valley from Building a New Global Underclass*. Boston: Houghton Mifflin Harcourt, 2019.

Greenfield, Adam. *Radical Technologies: The Design of Everyday Life*. London: Verso, 2017.

Gribbin, John. "Japanese Knock Spots Off Unmanned Factories." *New Scientist*, January 3, 1980.

Grierson, Jamie. "Amazon 'Regime' Making British Staff Physically and Mentally Ill, Says Union." *The Guardian*, August 18, 2015. http://www.theguardian.com/technology/2015/aug/18/amazon-regime-making-british-staff-physically-and-mentally-ill-says-union.

Gu, Xin, and Pip Shea. "Fabbing the Chinese Maker Identity." In *The Critical Makers Reader: (Un)Learning Technology*, edited by Loes Bogers and Letizia Chiappini, 269–77. Amsterdam: Institute for Network Cultures, 2019.

Gue, Kevin R. "The Human-Centric Warehouse." In *Progress in Material Handling Research*, 12. Milwaukee, 2010.

Guendelsberger, Emily. *On The Clock: What Low-Wage Work Did to Me and How It Drives America Insane*. New York: Little, Brown, 2020.

Gui, Xuemei. "Crisis Is a Test of Our Spiritual Strength." *Alibaba Tech*, August 13,

2020. https://alibabatech.medium.com/crisis-is-a-test-of-our-spiritual-strength-1270d8c6d85c.

Gurumurthy, Anita. "A Feminist Future of Work in the Post-Pandemic Moment: A New Social Contract as If Women Matter." Bengaluru: Feminist Digital Justice, April 2020. https://dawnnet.org/wp-content/uploads/2020/05/130520_DJP-Issue-No.-3_Final.pdf.

Gutelius, Beth, and Nik Theodore. "The Future of Warehouse Work: Technological Change in the U.S. Logistics Industry." Berkeley: UC Berkeley Center for Labor Research and Education, October 2019.

Han, Byung-Chul. *Psychopolitics: Neoliberalism and New Technologies of Power*. London: Verso Books, 2017.

———. *Shanzhai: Deconstruction in Chinese*. Cambridge, MA: MIT Press, 2017.

Harari, Yuval Noah. "How to Survive the 21st Century." Presented at the World Economic Forum, Davos, January 24, 2020. https://www.youtube.com/watch?v=eOsKFOrW5h8.

Haraway, Donna. *Simians, Cyborgs, and Women: The Reinvention of Nature*. New York: Routledge, 1990.

Harney, Stefano, and Fred Moten. *The Undercommons: Fugitive Planning and Black Study*. New York: Minor Compositions, 2013.

Harper, Douglas. "Automatic." *Online Etymology Dictionary*, 2021. https://www.etymonline.com/word/automatic.

Harwell, Drew, and Eva Dou. "Huawei Tested AI Software That Could Recognize Uighur Minorities and Alert Police, Report Says." *Washington Post*, December 9, 2020. https://www.washingtonpost.com/technology/2020/12/08/huawei-tested-ai-software-that-could-recognize-uighur-minorities-alert-police-report-says/.

Hatt, Beth. "Racializing Smartness." *Race Ethnicity and Education* 19, no. 6 (2016): 1141–48. https://doi.org/10.1080/13613324.2016.1168537.

Hawksworth, John, Richard Berriman, and Saloni Goel. "Will Robots Really Steal Our Jobs?" PricewaterhouseCoopers, 2018.

Heater, Brian. "Amazon Acquires Autonomous Warehouse Robotics Startup Canvas Technology." *TechCrunch* (blog), April 11, 2019. https://social.techcrunch.com/2019/04/10/amazon-acquires-autonomous-warehouse-robotics-startup-canvas-technology/.

———. "These Are the Robots That Help You Get Your Amazon Packages on Time." *TechCrunch* (blog), March 18, 2019. https://social.techcrunch.com/2019/03/17/these-are-the-robots-that-help-you-get-your-amazon-packages-on-time/.

Hegewisch, Ariane, Chandra Childers, and Heidi Hartmann. "Women, Automation and the Future of Work." Washington, DC: Institute for Women's Policy Research, 2019. https://iwpr.org/wp-content/uploads/2020/08/C476_Automation-and-Future-of-Work.pdf.

HelloMya. "I Had My First Amazonian Mental Breakdown at Work Last Night." reddit, December 2020. https://www.reddit.com/r/FASCAmazon/comments/jmla6z/i_had_my_first_amazonian_mental_breakdown_at_work/.

Hester, Helen. "Technically Female: Women, Machines, and Hyperemployment." *Salvage* (blog), August 8, 2016. https://salvage.zone/in-print/technically-female-women-machines-and-hyperemployment/.

Hester, Helen, and Nick Srnicek. "The Crisis of Social Reproduction and the End of Work." *OpenMind* (blog), March 2018. https://www.bbvaopenmind.com/en/articles/the-crisis-of-social-reproduction-and-the-end-of-work/.

Hicks, Mar. *Programmed Inequality: How Britain Discarded Women Technologists and Lost Its Edge in Computing.* Cambridge, MA: MIT Press, 2018. https://mitpress.mit.edu/books/programmed-inequality.

Hilliard, Constance. *Straightening the Bell Curve: How Stereotypes about Black Masculinity Drive Research on Race and Intelligence.* Washington, DC: Potomac Books, 2012.

Holusha, John. "General Motors: A Giant in Transition." *New York Times*, November 14, 1982. https://www.nytimes.com/1982/11/14/magazine/general-motors-a-giant-in-transition.html.

Hopkins, John. "Amazon Is No Ally in the Fight for Racial Justice. Interview by Alex Press, June 1, 2020." *Jacobin*, June 2020. https://jacobinmag.com/2020/06/amazon-racial-justice-worker-organizing-union.

Hui, Yuk. "Designing Tomorrow's Intelligence." Presented at the Belief in AI, Dubai, December 12, 2020. https://www.youtube.com/watch?v=9OodZPPFPMY.

———. *The Question Concerning Technology in China: An Essay in Cosmotechnics.* Cambridge, MA: MIT Press, 2019.

Human Rights Watch. "China: Big Data Program Targets Xinjiang's Muslims." Human Rights Watch, December 9, 2020. https://www.hrw.org/news/2020/12/09/china-big-data-program-targets-xinjiangs-muslims.

Hunt, Jami. "What Are Some Grocery Store Self-Checkout Hacks and Tricks?" Quora, January 10, 2020. https://www.quora.com/What-are-some-grocery-store-self-checkout-hacks-and-tricks.

Ingrassia, Paul, and Joseph B. White. *Comeback : The Fall and Rise of the American Automobile Industry.* New York: Simon & Schuster, 1995.

International Labour Organization. "World Employment and Social Outlook: The Role of Digital Labour Platforms in Transforming the World of Work." Geneva: International Labour Organization, 2021.

IPVM. "Huawei/Megvii Uyghur Alarms." IPVM, December 8, 2020. https://ipvm.com/reports/huawei-megvii-uygur.

Jäger, Georg, Laura S. Zilian, Christian Hofer, and Manfred Füllsack. "Crowdworking: Working with or against the Crowd?" *Journal of Economic Interaction and Coordination* 14, no. 4 (2019): 761–88. https://doi.org/10.1007/s11403-019-00266-1.

Joint Economic Committee. *Automation and Technological Change: Report of the Subcommittee on Economic Stabilization to the Joint Committee on the Economic Report, Congress of the United States.* Washington, DC: U.S. Government Printing Office, 1955.

Jünger, Friedrich Georg. *The Failure of Technology.* 1949. Reprint, Washington, DC: Regnery Gateway, 1990.

Kadir, Bzhwen A., Ole Broberg, and Carolina Souza da Conceição. "Designing Human-Robot Collaborations in Industry 4.0: Explorative Case Studies." 2018. https://doi.org/10.21278/idc.2018.0319.

Kaltman, Blaine. *Under the Heel of the Dragon: Islam, Racism, Crime, and the Uighur in China.* Athens: Ohio University Press, 2007.

Kantor, Jodi, and David Streitfeld. "Inside Amazon: Wrestling Big Ideas in a Bruising Workplace." *New York Times,* August 15, 2015. https://www.nytimes.com/2015/08/16/technology/inside-amazon-wrestling-big-ideas-in-a-bruising-workplace.html.

Kaplan, Marcia. "Alibaba's 2020 Singles Day Breaks Record, Attracts Luxury Brands." *Practical Ecommerce* (blog), November 15, 2020. https://www.practicalecommerce.com/alibabas-2020-singles-day-breaks-record-attracts-luxury-brands.

Kelly, Kevin. *The Inevitable: Understanding the 12 Technological Forces That Will Shape Our Future.* New York: Penguin Books, 2017.

———. *What Technology Wants.* New York: Penguin Books, 2011.

Kember, Sarah. "Notes towards a Feminist Futurist Manifesto." Edited by Kim Sawchuck and Carol Stabile. *Ada: A Journal of Gender New Media & Technology,* no. 1 (November 2012). https://adanewmedia.org/2012/11/issue1-kember/.

Keynes, John Maynard. "Economic Possibilities for Our Grandchildren." Madrid, June 1930. https://doi.org/10.1007/978-1-349-59072-8_25.

Khalid, Asma. "A Dirty Word in the U.S., 'Automation' Is a Buzzword in China," November 20, 2017. https://www.wbur.org/bostonomix/2017/11/20/china-automation.

Khatchadourian, Raffi. "Surviving the Crackdown in Xinjiang." *New Yorker,* April 12, 2021. https://www.newyorker.com/magazine/2021/04/12/surviving-the-crackdown-in-xinjiang.

Kong, Xiang, Hao Luo, George Huang, and Xuan Yang. "Industrial Wearable System: The Human-Centric Empowering Technology in Industry 4.0." *Journal of Intelligent Manufacturing* 30 (April 1, 2018): 1–17. https://doi.org/10.1007/s10845-018-1416-9.

Last Mile Prophets. *Cainiao's New Post Office Network Cainiao Post.* 2020. https://www.youtube.com/watch?v=7kUUIrsqIKc&t=249s.

Lawder, David, and Susan Heavey. "U.S. Blacklists China's Huawei as Trade Dispute Clouds Global Outlook." *Reuters*, May 16, 2019. https://www.reuters.com/article/us-usa-trade-china-idUSKCN1SL2DI.

Lecher, Colin. "How Amazon Automatically Tracks and Fires Warehouse Workers for 'Productivity.'" *Verge*, April 25, 2019. https://www.theverge.com/2019/4/25/18516004/amazon-warehouse-fulfillment-centers-productivity-firing-terminations.

Lee, Kai-Fu. *AI Superpowers: China, Silicon Valley, and the New World Order*. Boston: Houghton Mifflin Harcourt, 2018.

Leetaru, Kalev. "Is a Fragmented Internet Inevitable?" *Forbes*, April 13, 2019. https://www.forbes.com/sites/kalevleetaru/2019/04/13/is-a-fragmented-internet-inevitable/#74acd692223c.

Leibold, James. "Surveillance in China's Xinjiang Region: Ethnic Sorting, Coercion, and Inducement." *Journal of Contemporary China* 29, no. 121 (2020): 46–60.

Leontief, Wassily, and Faye Duchin. *The Future Impact of Automation on Workers*. New York: Oxford University Press, 1986.

Leung, Angela Ka-yee, and Dov Cohen. "Within- and Between-Culture Variation: Individual Differences and the Cultural Logics of Honor, Face, and Dignity Cultures." *Journal of Personality and Social Psychology* 100 (2011): 507–26. https://doi.org/10.1037/a0022151.

Lewig, Kerry, and Sara McLean. "Caring for Our Frontline Child Protection Workforce." Canberra: Child Family Community Australia, December 14, 2016. https://aifs.gov.au/cfca/publications/caring-our-frontline-child-protection-workforce.

Li, Fei-Fei, and John Etchemendy. "We Need a National Vision for AI." Stanford HAI, October 22, 2019. https://hai.stanford.edu/blog/we-need-national-vision-ai.

Lieber, Chavie. "Emergency Calls Placed from Amazon Warehouses Depict Enormous Pressure Put on Workers." *Vox*, March 11, 2019. https://www.vox.com/the-goods/2019/3/11/18260472/amazon-warehouse-workers-911-calls-suicide.

Lim, Louisa. "China: Re-Engineering the Uighur," November 7, 2018. https://www.lowyinstitute.org/the-interpreter/China-re-engineer-uighur.

Lin, Liza, and Shan Li. "Chinese Citizens Must Scan Their Faces to Register for New Mobile-Phone Service." *Wall Street Journal*, December 2, 2019. https://www.wsj.com/articles/chinese-citizens-must-scan-their-faces-to-register-for-new-mobile-phone-service-11575294169.

Linder, Natan. "Are We Inching toward the Lights Out Factory?" *Forbes*, June 17, 2020. https://www.forbes.com/sites/natanlinder/2020/06/17/are-we-inching-toward-the-lights-out-factory/.

Lindtner, Silvia, Anna Greenspan, and David Li. "Designed in Shenzhen: Shanzhai Manufacturing and Maker Entrepreneurs." *Aarhus Series on Human Centered Computing* 1, no. 1 (2015): 1–12. https://doi.org/10.7146/aahcc.v1i1.21265.

Liu, Xiulong, Jiannong Cao, Yanni Yang, and Shan Jiang. "CPS-Based Smart Ware-house for Industry 4.0: A Survey of the Underlying Technologies." *Computers* 7, no. 13 (2018): 1–17. https://doi.org/10.3390/computers7010013.

Long, Qiao. "Xinjiang Cell Phone Users Forced to Register with Real Names." Radio Free Asia, April 30, 2013. https://www.rfa.org/english/news/uyghur/register-04302013134824.html.

Lucas, Louise, Alice Woodhouse, and Emily Feng. "Alibaba Fights with Courier for Control of Customer Data." *Financial Times*, June 2, 2017. http://search.proquest.com/docview/1915368769/citation/E48419DD965A4677PQ/1.

Mahroof, Kamran. "A Human-Centric Perspective Exploring the Readiness towards Smart Warehousing: The Case of a Large Retail Distribution Warehouse." *International Journal of Information Management* 45 (2019): 176–90. https://doi.org/10.1016/j.ijinfomgt.2018.11.008.

Marshall, Margaret Wiley. "'Automation' Today and in 1662." *American Speech* 32, no. 2 (1957): 149–151. https://doi.org/10.2307/453032.

Marx, Karl. *Capital: A Critique of Political Economy*. Translated by Ben Fowkes. London: Penguin Books, 2004.

Mateescu, Alexandra, and Madeleine Elish. "AI in Context: The Labor of Integrating New Technologies." New York: Data and Society, January 2019.

Mattes, Eva, and Franco Mattes. *Dark Content*. 2015. Installation and video. https://0100101110101101.org/dark-content/.

McAlevey, Jane, and Bob Ostertag. *Raising Expectations (and Raising Hell): My Decade Fighting for the Labor Movement*. London: Verso, 2014.

McGirt, Ellen. "Amazon's Echo Device Chief on the Risk of Alexa's Many Rewards." *Fortune*, July 17, 2018. http://fortune.com/2018/07/16/amazon-alexa-echo-toni-reid/.

McKinsey Global Institute. "Automation, Jobs, and the Future of Work." McKinsey and Company, December 1, 2014. https://www.mckinsey.com/featured-insights/employment-and-growth/automation-jobs-and-the-future-of-work.

Melamed, Jodi. "Racial Capitalism." *Critical Ethnic Studies* 1, no. 1 (2015): 76–85.

Merchant, Brian. "Why Self-Checkout Is and Has Always Been the Worst." Gizmodo, March 7, 2019. https://gizmodo.com/why-self-checkout-is-and-has-always-been-the-worst-1833106695.

Mezzadra, Sandro, and Brett Neilson. "Extraction, Logistics, Finance: Global Crisis and the Politics of Operations." *Radical Philosophy* 178 (March/April 2013): 8–18.

Michalski, Jerry. "AI, Robotics, and the Future of Jobs." *Pew Research Center: Internet, Science & Tech* (blog), August 6, 2014. https://www.pewresearch.org/internet/2014/08/06/future-of-jobs/.

Mindell, David A. *Our Robots, Ourselves: Robotics and the Myths of Autonomy*. New York: Viking Press, 2015.

Moore, Jason. *Capitalism in the Web of Life: Ecology and the Accumulation of Capital.* London: Verso, 2015.

———. "Sugar and the Expansion of the Early Modern World-Economy: Commodity Frontiers, Ecological Transformation, and Industrialization." *Review (Fernand Braudel Center)* (2000): 409–33.

Moore-Colyer, Roland. "Amazon's Human Workers Are Safe for Now, but the Tide of Automation Is Rising." IT PRO, May 3, 2019. https://www.itpro.co.uk/business-strategy/33565/amazons-human-workers-are-safe-for-now-but-the-tide-of-automation-is-rising.

Moradi, Pegah. "Race, Ethnicity, and the Future of Work." Bachelor of arts thesis, Cornell University, 2019. https://doi.org/10.31235/osf.io/e37cu.

Morrison, Toni. *Playing in the Dark: Whiteness and the Literary Imagination.* New York: Vintage, 2007.

Morris-Suzuki, Tessa. "Robots and Capitalism." *New Left Review* 147, no. 1 (1984): 109–21.

Mosco, Vincent. *The Digital Sublime: Myth, Power, and Cyberspace.* Cambridge, MA: MIT Press, 2005.

Mozur, Paul. "One Month, 500,000 Face Scans: How China Is Using A.I. to Profile a Minority." *New York Times*, April 14, 2019. https://www.nytimes.com/2019/04/14/technology/china-surveillance-artificial-intelligence-racial-profiling.html.

Mumford, Lewis. *The Myth of the Machine: Technics and Human Development.* New York: Harcourt, 1967.

Munn, Luke. *Ferocious Logics: Unmaking the Algorithm.* Lüneburg: meson press, 2018.

———. "Red Territory: Forging Infrastructural Power." *Territory, Politics, Governance*, October 20, 2020, 1–20. https://doi.org/10.1080/21622671.2020.1805353.

Musk, Elon. "@timkhiggins Yes, Excessive Automation at Tesla Was a Mistake. To Be Precise, My Mistake. Humans Are Underrated." Tweet. *@elonmusk* (blog), April 13, 2018. https://twitter.com/elonmusk/status/984882630947753984.

NAMI. "Law Enforcement." National Alliance on Mental Illness, 2020. https://www.nami.org/Advocacy/Crisis-Intervention/Law-Enforcement.

New China TV. *The Rise of Modern Farming in Xinjiang's Cotton Fields.* Xinjiang, 2020. https://www.youtube.com/watch?v=nJovoOqiXAU.

New York Times. "Man Devoured by His Machines." *New York Times*, October 2, 1921. https://timesmachine.nytimes.com/timesmachine/1921/10/02/107026085.pdf.

Newlands, Gemma, and Christoph Lutz. "Crowdwork and the Mobile Underclass: Barriers to Participation in India and the United States." *New Media and Society* 19 (2020): 1868–86. https://doi.org/10.1177/1461444820901847.

Newton, Casey. "Facebook Will Pay $52 Million in Settlement with Modera-

tors Who Developed PTSD on the Job." *Verge*, May 12, 2020. https://www.
theverge.com/2020/5/12/21255870/facebook-content-moderator-settlement
-scola-ptsd-mental-health.

Noble, David F. *Forces of Production: A Social History of Industrial Automation*. New
Brunswick, NJ: Transaction, 2011.

O'Connor, Sarah. "Amazon Unpacked." *Financial Times*, February 9, 2013.

Oliver, Melvin, and Thomas Shapiro. *Black Wealth/White Wealth: A New Perspective
on Racial Inequality*. New York: Routledge, 2006.

Oppenheimer, Andres. *The Robots Are Coming!: The Future of Jobs in the Age of Automation*. Translated by Ezra E. Fitz. New York: Vintage, 2019.

Pacey, Arnold. *The Culture of Technology*. Cambridge, MA: MIT Press, 1983.

Parisi, Luciana. "Improper Commonness: The Surrogate Economy of Artificial Intelligence." In *Hyperemployment: Post-Work, Online Labour and Automation*, edited by !Mediengruppe Bitnik, Silvio Lorusso, Luciana Parisi, Domenico Quar-
anta, and Felix Stalder, 64–77. Rome: NERO Editions, 2020.

Pelegrin, Marc. "An Argument against Automation." *Technology in Society* 2, no. 4
(1980): 433–47. https://doi.org/10.1016/0160-791X(80)90005-6.

Pellow, David, and Lisa Sun-Hee Park. *The Silicon Valley of Dreams: Environmental Injustice, Immigrant Workers, and the High-Tech Global Economy*. New York:
New York University Press, 2003.

Phillips, Tom. "China Orders Hundreds of Thousands of Private Cars to Have GPS
Trackers Installed for Monitoring." *Guardian*, February 21, 2017. http://www.
theguardian.com/world/2017/feb/21/china-orders-gps-tracking-of-every-car-in
-troubled-region.

Plant, Sadie. "Genderquake." In *Zeroes + Ones: Digital Women + the New Technoculture*, 37–44. New York: Doubleday, 1997.

Plous, Scott, and Tyrone Williams. "Racial Stereotypes from the Days of American
Slavery: A Continuing Legacy." *Journal of Applied Social Psychology* 25, no. 9
(1995): 795–817.

Price, Anne. "Black People vs Robots: Reparations and Workers Rights in the Age of
Automation." Presented at Data for Black Lives, Boston, January 15, 2019. https://
www.youtube.com/watch?v=1uOt3GriWAI.

Purington, Amanda, Jessie G. Taft, Shruti Sannon, Natalya N. Bazarova, and Samuel
Hardman Taylor. "Alexa Is My New BFF: Social Roles, User Satisfaction, and
Personification of the Amazon Echo." In *Proceedings of the 2017 CHI Conference
Extended Abstracts on Human Factors in Computing Systems*, 2853–59. New York:
ACM, 2017.

浮图. "河南女工去新疆摘棉花挣钱 几天后就后悔，坐棉花地痛哭. " [QQ.com]. [A
female worker from Henan went to Xinjiang to pick cotton to make money. A
few days later, she regretted it and cried bitterly while sitting on the cotton field].

Quanguo, Chen. "陈全国：筑起反恐维稳的铜墙铁壁 确保新疆社会大局和谐稳定_央广网" [Build a copper wall and iron wall to fight terrorism and maintain stability to ensure the overall harmony and stability of Xinjiang's society]. August 17, 2017. http://news.cnr.cn/native/city/20170819/t20170819_523908751.shtml.

Raghav, Krish. *Bullshit Jobs in China.* Beijing: Krish Raghav, 2021. https://gumroad.com/d/6327bf648c52045071f78af386069d88.

Ramzy, Austin, and Chris Buckley. "'Absolutely No Mercy': Leaked Files Expose How China Organized Mass Detentions of Muslims." *New York Times*, November 16, 2019. https://www.nytimes.com/interactive/2019/11/16/world/asia/china-xinjiang-documents.html.

Ransome, Paul. *Work, Consumption and Culture: Affluence and Social Change in the Twenty-First Century.* London: Sage, 2005.

Ray, Tiernan. "OpenAI's Gigantic GPT-3 Hints at the Limits of Language Models for AI." ZDNet, June 1, 2020. https://www.zdnet.com/article/openais-gigantic-gpt-3-hints-at-the-limits-of-language-models-for-ai/.

Recruitme. "Machine Minder Rockit Packhouse." LinkedIn, March 15, 2021. https://nz.linkedin.com/jobs/view/machine-minder-rockit-packhouse-recruitme-at-recruitme-2492042361.

Rhee, Jennifer. *The Robotic Imaginary: The Human and the Price of Dehumanized Labor.* Minneapolis: University of Minnesota Press, 2018.

Rifkin, Jeremy. *The End of Work: The Decline of the Global Labor Force and the Dawn of the Post-Market Era.* New York: Tarcher, 1996.

Rivera, Alex. *Sleep Dealer.* Maya Entertainment, 2008.

Roberts, Carys, Henry Parkes, Rachel Statham, and Lesley Rankin. "The Future Is Ours: Women, Automation and Equality." London: Institute for Public Policy Research, July 2019.

Roberts, Huw, Josh Cowls, Jessica Morley, Mariarosaria Taddeo, Vincent Wang, and Luciano Floridi. "The Chinese Approach to Artificial Intelligence: An Analysis of Policy, Ethics, and Regulation." *AI and Society*, June 17, 2020. https://doi.org/10.1007/s00146-020-00992-2.

Roberts, Sarah. Content Moderation of Social Media, April 12, 2021. https://www.youtube.com/watch?v=ao7fTW6B150&t=2052s.

——— *Behind the Screen: Content Moderation in the Shadows of Social Media.* New Haven: Yale University Press, 2019.

Roberts, Sean. *The War on the Uyghurs: China's Internal Campaign against a Muslim Minority.* Princeton, NJ: Princeton University Press, 2020. https://aaaaarg.fail/thing/5f6da2999ff37c5b0e2e81cc.

Robinson, Cedric J. *Black Marxism: The Making of the Black Radical Tradition.* Durham: University of North Carolina Press, 2000.

Robinson, Kim Stanley. *New York 2140.* London: Hachette UK, 2017.

Roediger, David R. *The Wages of Whiteness: Race and the Making of the American Working Class*. London: Verso, 2007.

Romero, David, Peter Bernus, Ovidiu Noran, Johan Stahre, and Åsa Fast-Berglund. "The Operator 4.0: Human Cyber-Physical Systems and Adaptive Automation Towards Human-Automation Symbiosis Work Systems." In *Advances in Production Management Systems: Initiatives for a Sustainable World*, edited by Irenilza Nääs, Oduvaldo Vendrametto, João Mendes Reis, Rodrigo Franco Gonçalves, Márcia Terra Silva, Gregor von Cieminski, and Dimitris Kiritsis, 677–86. Cham: Springer International, 2016. https://doi.org/10.1007/978-3-319-51133-7_80.

Rose, Sarah. *No Right to Be Idle: The Invention of Disability, 1840s–1930s*. Chapel Hill, NC: UNC Press, 2017.

Rossiter, Ned. *Software, Infrastructure, Labor: A Media Theory of Logistical Nightmares*. New York: Routledge, 2017.

Rothschild, Joan. *Women, Technology, and Innovation*. Oxford: Pergamon Press, 1982.

Rowanissupreme. "The Mental Health at Amazon Seems Debilitating." reddit, February 2021. https://www.reddit.com/r/FASCAmazon/comments/lt3w59/the_mental_health_at_amazon_seems_debilitating/.

Rushkoff, Douglas. "Silicon Valley's Push for Universal Basic Income Is—Surprise!—Totally Self-Serving." *Los Angeles Times*, July 21, 2017. https://www.latimes.com/opinion/op-ed/la-oe-rushkoff-universal-basic-income-silicon-valley-20170721-story.html.

Russell, Matt. "Automated Picking Bins Set to Help New Zealand Growers Save Labour Costs." Fresh Plaza, June 10, 2020. https://www.freshplaza.com/article/9224099/automated-picking-bins-set-to-help-new-zealand-growers-save-labour-costs/.

Sadowski, Jathan. "Why Silicon Valley Is Embracing Universal Basic Income." *Guardian*, June 22, 2016. http://www.theguardian.com/technology/2016/jun/22/silicon-valley-universal-basic-income-y-combinator.

Sampath, Meera, and Pramod Khargonekar. "Socially Responsible Automation: A Framework for Shaping the Future." *NAE Website* 48, no. 4 (n.d.): 45–52.

Scheiber, Noam. "Inside an Amazon Warehouse, Robots' Ways Rub Off on Humans." *New York Times*, July 3, 2019. https://www.nytimes.com/2019/07/03/business/economy/amazon-warehouse-labor-robots.html.

Scholz, Trebor. "Platform Cooperativism." New York: Rosa Luxemburg Foundation, 2016. http://www.rosalux-nyc.org/wp-content/files_mf/scholz_platformcoop_5.9.2016.pdf.

Schwartz, Jeff. "Yes, the Robots Are Coming: Jobs in the Era of Humans and Machines." *Wired*, May 24, 2018. https://www.wired.com/wiredinsider/2018/05/yes-the-robots-are-coming-jobs-in-the-era-of-humans-and-machines/.

Schwartz, Oscar. "Untold History of AI: Invisible Women Programmed America's

First Electronic Computer." *IEEE Spectrum: Technology, Engineering, and Science News*, May 25, 2019. https://spectrum.ieee.org/tech-talk/tech-history/dawn-of-electronics/untold-history-of-ai-invisible-woman-programmed-americas-first-electronic-computer.

Schwedel, Heather. "What Life with an Amazon Echo Is Like after the Novelty Wears Off." *Slate Magazine*, December 26, 2017. https://slate.com/technology/2017/12/what-life-with-an-amazon-echo-is-like-after-the-novelty-wears-off.html.

Seppelt, Bobbie D., and John D Lee. "Keeping the Driver in the Loop: Dynamic Feedback to Support Appropriate Use of Imperfect Vehicle Control Automation." *International Journal of Human-Computer Studies* 125 (2019): 66–80.

Seta, Gabriele de. "Into the Red Stack." *Hong Kong Review of Books*, 香港書評, April 17, 2018. https://hkrbooks.com/2018/04/17/into-the-red-stack/.

Simon, Herbert. *The Shape of Automation for Men and Management*. New York: Harper & Row, 1965.

Simon, Matt. "Inside the Amazon Warehouse Where Humans and Machines Become One." *Wired*. Accessed September 16, 2020. https://www.wired.com/story/amazon-warehouse-robots/.

Singer, Peter Warren. *Wired for War: The Robotics Revolution and Conflict in the 21st Century*. London: Penguin, 2009.

Smiley, Erica. "Black People vs Robots: Reparations and Workers Rights in the Age of Automation." Presented at Data for Black Lives, Boston, January 15, 2019. https://www.youtube.com/watch?v=1uOt3GriWAI.

Smith, Jason. *Smart Machines and Service Work: Automation in an Age of Stagnation*. London: Reaktion Books, 2020.

Sparrow, Robert. "Robotics Has a Race Problem." *Science, Technology, and Human Values* 45, no. 3 (2020): 538–60.

Spatharou, Angela, Solveigh Hieronimus, and Jonathan Jenkins. "Transforming Healthcare with AI: The Impact on the Workforce and Organizations." London: McKinsey, March 10, 2020. https://www.mckinsey.com/industries/healthcare-systems-and-services/our-insights/transforming-healthcare-with-ai.

Spitznagel, Eric. "Inside the Hellish Workday of an Amazon Warehouse Employee." *New York Post* (blog), July 13, 2019. https://nypost.com/2019/07/13/inside-the-hellish-workday-of-an-amazon-warehouse-employee/.

Srnicek, Nick, and Alex Williams. *Inventing the Future: Postcapitalism and a World without Work*. London: Verso Books, 2015.

Stark, Louis. "Does Machine Displace Men in the Long Run?" *New York Times*, February 25, 1940. https://www.nytimes.com/1940/02/25/archives/does-machine-displace-men-in-the-long-run-new-studies-cited-as-old.html.

Sudworth, John. "China's 'Tainted' Cotton." *BBC News*, December 2020. https://bbc.co.uk/news/extra/nzog306v8c/china-tainted-cotton.

———. "Inside China's Scheme to Transfer Uighurs into Work: 'If the Others Go I'll Go.'" RNZ, March 3, 2021. https://www.rnz.co.nz/news/world/437597/ inside-china-s-scheme-to-transfer-uighurs-into-work-if-the-others-go-i-ll-go.

Susskind, Daniel. *A World without Work: Technology, Automation and How We Should Respond*. London: Penguin UK, 2020.

SystematicApproach. "Wife Sucks the Fun out of HA." Reddit Post. *R/Homeautomation*, March 13, 2017. www.reddit.com/r/homeautomation/comments/5z88ok/ wife_sucks_the_fun_out_of_ha/.

Tao, Tao, and Xiaofa Shi. "Logistic Supply and Demand Analysis on Online Shopping of Double Eleven Day." In *COTA International Conference of Transportation Professionals 2017*, 1453–59. Shanghai: American Society of Civil Engineers, 2018. https://doi.org/10.1061/9780784480915.154.

Taylor, Emmeline. "Supermarket Self-Checkouts and Retail Theft: The Curious Case of the SWIPERS." *Criminology and Criminal Justice* 16, no. 5 (2016): 552–67. https://doi.org/10.1177/1748895816643353.

Taylor, Frederick Winslow. *The Principles of Scientific Management*. New York: Harper, 1913.

Te Mana Raraunga. "What We Do." Te Mana Raraunga, 2021. https://www.temanararaunga.maori.nz/kaupapa.

TechCrunch. *Canvas' Robot Cart Could Change How Factories Work*, 2017. https:// www.youtube.com/watch?v=e9PsHFgWaxc.

Teplitzky, Alex. "Remote Warfare, Automation, and Digital Labor: Why Alex Rivera's 2008 Film SLEEP DEALER Is Still Relevant." *Creative Capital*, May 14, 2019. https://creative-capital.org/2019/05/14/remote-warfare-automation-and-digital-labor-why-alex-riveras-2008-film-sleep-dealer-is-still-relevant/.

Time. "The Automation Jobless." *Time*, February 24, 1961.

TSC Foods. "Machine Minder." TSC Foods, March 1, 2021. http://www.tscfoods. com/careers/machine-minder/.

Tsin, Ong Peng. "Betting Big on Technification in South-East Asia." *Business Times*, December 16, 2019. https://www.businesstimes.com.sg/garage/ betting-big-on-technification-in-south-east-asia.

Turk, Victoria. "Home Invasion." *New Scientist* 232, no. 3104 (2016): 16–17. https:// doi.org/10.1016/S0262-4079(16)32318-1.

Turner-Lee, Nicol. "Black People vs Robots: Reparations and Workers Rights in the Age of Automation." Presented at Data for Black Lives, Boston, January 15, 2019. https://www.youtube.com/watch?v=1uOt3GriWAI.

Vonnegut, Kurt. *Player Piano*. New York: Dial Press, 1952.

Wajcman, Judy. "Automation: Is It Really Different This Time?" *British Journal of Sociology* 68, no. 1 2017): 119–27. https://doi.org/10.1111/1468-4446.12239.

———. *Pressed for Time: The Acceleration of Life in Digital Capitalism*. Chicago: University of Chicago Press, 2015.

Wang, Cunrui, Qingling Zhang, Wanquan Liu, Yu Liu, and Lixin Miao. "Facial Feature Discovery for Ethnicity Recognition." *Wiley Interdisciplinary Reviews: Data Mining and Knowledge Discovery* 9, no. 1 (2019): e1278. https://doi.org/10.1002/widm.1278.

Wang, Maya. "China's Algorithms of Repression." New York: Human Rights Watch, May 1, 2019. https://www.hrw.org/report/2019/05/01/chinas-algorithms-repression/reverse-engineering-xinjiang-police-mass.

Wee, Sui-Lee. "China Uses DNA to Track Its People, with the Help of American Expertise." *New York Times*, February 21, 2019. https://www.nytimes.com/2019/02/21/business/china-xinjiang-uighur-dna-thermo-fisher.html.

weepinwidow. "Amazon and Mental Health." reddit, January 2021. https://www.reddit.com/r/FASCAmazon/comments/k96ujq/amazon_and_mental_health/.

Wells, Jean Alice. "Automation and Women Workers." Washington, DC: U.S. Department of Labor, Wage and Labor Standards Administration, Women's Bureau, 1970.

Wesson, Robert W. "Materials Handling." *Industrial & Engineering Chemistry* 50, no. 3 (1958): 474–77.

Wilshaw, Paul. "The Ethical Automation Toolkit." Medium, May 25, 2019. https://towardsdatascience.com/the-ethical-automation-toolkit-f1fd4281534e.

Witt, P. R. "Automatic Storage and Retrieval System Control." In *Proceedings of the 1974 National Computer Conference and Exposition*, 593–611. New York: Association for Computing Machinery, 1974. https://doi.org/10.1145/1500175.1500292.

Wynter, Sylvia. "Unsettling the Coloniality of Being/Power/Truth/Freedom: Towards the Human, after Man, Its Overrepresentation—An Argument." *CR: The New Centennial Review* 3, no. 3 (2003): 257–337. https://doi.org/10.1353/ncr.2004.0015.

Xin, Chang. "The Police Region of Xinjiang: Checkpoints, Camps, and Fear," May 9, 2020. https://bitterwinter.org/the-police-region-of-xinjiang-checkpoints-camps-and-fear/.

Yang, Guobin. "A Chinese Internet? History, Practice, and Globalization." *Chinese Journal of Communication* 5, no. 1 (2012): 49–54. https://doi.org/10.1080/17544750.2011.647744.

Yang, Yuan. "Xinjiang Phone App Exposes How Chinese Police Monitor Uighur Muslims," May 1, 2019. https://www.ft.com/content/dfec4ac4-6bf5-11e9-80c7-60ee53e6681d.

Ye, Josh. "Data Spat between Home Delivery Majors Causes Huge Confusion." *South China Morning Post*, June 2, 2017. https://www.scmp.com/business/article/2096631/cainiao-sf-express-standoff-over-data-gumming-deliveries-chinas-online.

Yongzhou, Yang. "Why Cainiao's Special and How Its Elastic Scheduling System

Works." Alibaba Tech, January 2, 2020. https://www.alibabacloud.com/blog/why-cainiaos-special-and-how-its-elastic-scheduling-system-works_595702.

Zahn, Max, and Sharif Paget. "'Colony of Hell': 911 Calls from Inside Amazon Warehouses." *The Daily Beast*, March 11, 2019, sec. tech. https://www.thedailybeast.com/amazon-the-shocking-911-calls-from-inside-its-warehouses.

Zenz, Adrian. "Coercive Labor in Xinjiang: Labor Transfer and the Mobilization of Ethnic Minorities to Pick Cotton." Washington, DC: Newlines Institute for Strategy and Policy, December 2020.

Zheng, Dan, and Tatsuaki Kuroda. "The Impact of Economic Policy on Industrial Specialization and Regional Concentration of China's High-Tech Industries." *Annals of Regional Science* 50, no. 3 (2013): 771–90. https://doi.org/10.1007/s00168-012-0522-4.

Zimmerman, Jan. *The Technological Woman: Interfacing with Tomorrow*. ABC-CLIO, LLC, 1983.

Index